The Chemical Age

For Amy & Matt
& friendship

3/5/22

THE CHEMICAL AGE

How Chemists Fought Famine and Disease, Killed Millions, and Changed Our Relationship with the Earth

FRANK A. VON HIPPEL

The University of Chicago Press

Chicago and London

The University of Chicago Press, Chicago 60637
The University of Chicago Press, Ltd., London
© 2020 by The University of Chicago
All rights reserved. No part of this book may be used or reproduced in any
manner whatsoever without written permission, except in the case of brief
quotations in critical articles and reviews. For more information, contact
the University of Chicago Press, 1427 E. 60th St., Chicago, IL 60637.
Published 2020
Printed in the United States of America

29 28 27 26 25 24 23 22 21 20   2 3 4 5

ISBN-13: 978-0-226-69724-6   (cloth)
ISBN-13: 978-0-226-69738-3   (e-book)

DOI: https://doi.org/10.7208/chicago/9780226697383.001.0001

Library of Congress Cataloging-in-Publication Data

Names: Von Hippel, Frank A. (Frank Arthur), author.
Title: The chemical age : how chemists fought famine and disease, killed millions,
    and changed our relationship with the earth / Frank A. von Hippel.
Description: Chicago ; London : University of Chicago Press, 2020. |
    Includes bibliographical references and index.
Identifiers: LCCN 2019046215 | ISBN 9780226697246 (cloth) |
    ISBN 9780226697383 (ebook)
Subjects: LCSH: Pesticides—History. | Chemical weapons—History. |
    Pesticides—Environmental aspects. | Chemical industry—Environmental
    aspects. | Human ecology—History. | Environmentalism—History.
Classification: LCC TD196.P38 V66 2020 | DDC 632/.9509—dc23
LC record available at https://lccn.loc.gov/2019046215

For Cathy, and the next generation

Contents

Prologue

In 1921, the brilliant inventor Thomas Midgley Jr. discovered that he could eliminate knocking in internal combustion automobile engines and dramatically boost performance by adding tetraethyl lead to gasoline.[1] However, this resulted in the deposition of lead oxide in the engine, which damaged spark plugs and exhaust valves. To solve the problem, Midgley and his team experimented with scavenger compounds of chlorine and bromine that bind to lead during combustion, thereby releasing it as exhaust. In 1925, they chose ethylene dibromide as the ideal scavenger. Producing this scavenger in immense quantities required extracting bromine from the ocean, and Midgley quickly developed a suitable method.[2] The team called the resulting fuel, which they tinted red as a marketing ploy, ethyl gasoline.[1]

Midgley suffered lead poisoning during the development of ethyl gasoline, but after a break from research, he recovered.[2] Meanwhile, workers at three production facilities either died or developed psychosis.[3] Nevertheless, Standard Oil and DuPont produced the fuel on a large scale, and in the next half-century, drivers filled their cars with 25 trillion liters of leaded gasoline.[4] The resulting lead pollution released from tailpipes worldwide caused irreversible losses in intelligence and increases in impulsive and aggressive behavior in exposed children.[5,6] Scientists subsequently linked the neurological effects of airborne lead pollution to rising rates of delinquency, violent crime, and unwed pregnancy associated with exposed children becoming young adults.[7-15]

Another of Midgley's chemical innovations followed a similar trajectory. Refrigerants in the 1920s were toxic and had an unfortunate tendency to catch fire or explode.[16] Midgley and his team synthesized various compounds to find a replacement chemical that would be volatile, chemically inert, and nontoxic. The solution they found after merely three days of experiments in 1930 incorporated fluorine into a hydrocarbon to make dichlorodifluoromethane, the first chlorofluorocarbon, or CFC.[16,17] General Motors and DuPont marketed the compound as Freon. To demonstrate Freon's safety, Midgley inhaled some before an audience and then blew out a candle with his Freon-saturated breath.[17]

Freon and subsequent CFCs proved to be a regrettable replacement for prior refrigerants as they destroyed ozone in the stratosphere, which protects life on the planet from damaging ultraviolet radiation. Mario Molina and Frank Sherwood Rowland discovered this global threat of CFCs in 1974,[18] nearly half a century after Midgley first produced Freon. For their planet-saving work, Molina and Rowland received the 1995 Nobel Prize in Chemistry.[19] A year after Molina and Rowland discovered that CFCs destroy ozone, another scientist demonstrated that CFCs are also potent greenhouse gases contributing to global warming.[20]

Environmental historian J. R. McNeill declared that Midgley "had more impact on the atmosphere than any other single organism in earth history."[4] In his attempts to improve the world, Midgley managed to invent products that were responsible for causing neurological damage in countless children and for potentially making the Earth uninhabitable. But he did not live to see the downsides of his discoveries. Polio robbed him of his mobility in 1940.[17] Ever the inventor, he designed a machine to facilitate his entry and exit from bed; on November 2, 1944, at the age of fifty-five, Midgley suffocated in the ropes of his contraption.

The story of Thomas Midgley Jr. is part of a long history of chemists and chemical engineers who, when confronted by intractable problems, produced a startling array of products and applications—often resulting in serious unintended consequences. This book is about that history, and about the scientists who strove to halt famine, infectious diseases, and opposing armies with chemistry. Some of these scientists started with pure intentions, but slid into the darkest depths of depravity. In a disquieting number of cases, chemicals designed to prevent famine and plague were eventually wielded for evil, and chemicals designed to do evil were afterward employed to do good.

This book is also a story of human folly, prejudice, slavery, and genocide; of the scattering of ethnic groups and the destruction of nature; and of scientists scrambling to forge a world free of hunger and disease. It is the story of competition between scientists to be the first to make a discovery and of the occasional realization of the triviality of their rivalries in relation to the events of war unfolding around them. It is the tangled story of how chemicals link the histories of famine, plague, and war; how humanity's uneasy coexistence with pests and longstanding battles to exterminate them have shaped history; and how our relationship with pesticides eventually made possible a new era of ecological awareness.

This book explores the period of 1845 to 1964, with some meandering to both more recent times and back as far as 2700 BCE. The primary historical range is bounded on the early end by the Irish Potato Famine and on the late end by the firestorm over the publication of Rachel Carson's *Silent Spring*. The book starts with a tragedy that led scientists on an urgent mission to prevent famine with chemicals, and then traces the centuries during which epidemics tore through humanity's ranks. The subsequent discovery of disease pathogens and their transmission by animals presented a means to end misery: pesticides were quickly developed to kill these animal vectors. Inevitably, scientists discovered that many of these chemicals could be weaponized, and the world experienced waves of chaos as modern warfare swept over national boundaries. The complex, two-way relationship between pesticides and chemical weapons solidified, and chemical companies accumulated wealth and power. The use of chemicals in war led to a rush to harness the power of chemistry against pests during times of peace. All the while, chemical discoveries ratcheted the world into new realities in which hunger and disease retreated into smaller geographic regions, chemical weapons became accessible to more belligerents, and persistent pollutants contaminated even the most remote habitats on the planet. The book ends with the dawn of society's realization that pesticides often eat away the very fabric of human and ecological health, sometimes leading to human tragedy and driving species to extinction.

Who were the scientists battling pests and opposing armies with chemistry? The science essential for the synthesis of new chemicals developed in the cultural context of nineteenth- and twentieth-century imperial ambitions. The scientists, embedded in national conflicts, and shaped by adulation and fame, took extraordinary risks on the path to extraordinary discoveries. The hazards of these novel chemicals, for both human and environmental health, were in some cases discovered through deliberate experimentation on humans, while in other cases they were stumbled upon by keen observers.

The most important of these observers was Rachel Carson. Her struggles to alert the world to chemical risks not only ushered in the environmental movement, but also revealed the extent to which human health is dependent upon functioning ecosystems. Her work encouraged holistic thinking about humanity's place in the natural world. These ideas, so important to our future, have their roots in the story of pesticides. This book is ultimately about the birth of these ideas, which occurred in the context of our long and tumultuous relationship with lethal chemicals. Here is the story of the people who dragged our world, for better or for worse, into the chemical age.

Author's Note

Citations: This book relies on primary source material, with documents extending back to the fourteenth century. Every fact in the book is cited. When a fact has no citation, it is drawn from the immediately preceding citation within the same paragraph. In addition to primary source material, this book also draws upon original translations of non-English sources and scholarly work that synthesized topical areas.

Units of measure: The units reported in this book are the units used in the primary source material.

Spelling: American English spelling is used throughout, except that quotations from historical sources are represented with their original spelling, reflecting in some cases changes to the English language, including word choice, and in other cases British spelling.

PART 1

Famine

[**1**]

Potato Blight

(1586–1883)

I have visited the wasted remnants of the once noble Red Man, on his reservation grounds in North America, and explored the "negro quarter" of the degraded and enslaved African, but never have I seen misery so intense, or physical degradation so complete, as among the dwellers in the bog-holes of Erris.—**James H. Tuke**, Autumn 1847[21]

Potatoes are the world's fourth most important crop, and for some nations they are the primary food source. But potatoes are vulnerable to an array of pests, which have caused sporadic and severe famine across time. The story of the potato and its pests is the story of the globalization of commerce, of famine and deadly disease outbreaks, and of the search for chemical agents to kill plant pathogens and the insects that consume potatoes or serve as vectors for diseases. Those chemical agents are pesticides, and pesticides are woven into the account of more than a century of progress in the battles against hunger and disease. Pesticides also are integral to the history of modern warfare and environmental destruction. A logical place to begin a history of pesticides is with a chronicle of the potato and the famine it brought to Ireland.

3

The history of potatoes begins with eight thousand years of cultivation in the Andes, where the indigenous people developed thousands of potato varieties.[22] Some Andean farmers raised as many as two hundred varieties on a single plot of land. In the sixteenth century, Spanish explorers brought the potato from the Inca empire to Spain and then to Florida, from where colonists took it to Virginia. From Virginia, the well-traveled potato returned to Europe. Sir Thomas Herriot, a companion of Sir Walter Raleigh, transported the potato to Great Britain in 1586.[23] A few years later, the famous botanist Gaspard Bauhin gave the potato its scientific name, *Solanum tuberosum*. *Solanum* is derived from the Latin meaning "soothing" or "quieting," though this tuber did not have a soothing future.

The first significant European cultivation of potatoes occurred in Ireland, near Cork, and then spread to farms on the Continent; however, the plant's membership in the family of the deadly nightshade (the Solanaceae) cast a shadow over its reputation.[23] Indeed, many people blamed the potato for leprosy and other ailments. Noble efforts eased the potato's path to acceptance, though that, too, was a rocky road. Sir Walter Raleigh managed to convince Queen Elizabeth I to allow the potato to grace the royal table, an effort that flopped. "Courtesy forbade the guests to refuse to partake of the new dish," wrote the authors of a 1906 history of potatoes, "but their dislike was so obvious, and so assiduously did they circulate tales regarding the poisonous nature of the tubers, that we do not read of the experiment being repeated."[24] The potato took hold in Ireland, but not in England—until 1663, when the Royal Society pushed for its general cultivation because of its value in times of famine.[23]

In France, the cultivation of potatoes was forbidden until the charismatic army pharmacist Antoine-Augustin Parmentier made it his task to advocate for them. Parmentier had eaten potatoes while a prisoner of war in Prussia.[25,26] After he returned to France, he convinced the Paris Faculty of Medicine to declare potatoes edible in 1772. Public acceptance was harder to achieve, so Parmentier used trickery to convince people that potatoes were safe to eat. He received the royal blessing of King Louis XVI to generate popular curiosity by stationing troops to guard his potato plot.[23] The troops were permitted to accept bribes from people wanting potatoes, and the soldiers withdrew at night to allow for theft.[25,26]

Parmentier also used his harvest to create feasts of potatoes. Men of influence, including Benjamin Franklin, were invited and convinced.[23] Louis XVI, festooned with a potato flower in his buttonhole, expanded the tuber's acceptance when he ordered large-scale cultivation. By 1813, the Central Society of Agriculture had collected well over one hundred potato varieties under cultivation in the French empire. The potato was known in French as *pomme de terre*, or "apple of the earth."

Potatoes became particularly important in Ireland, where they could be grown on lands unsuitable for other crops.[21] The Irish had long ago been pushed off the best lands by English landlords, who grazed cattle destined for the British market. But even on marginal lands, in bogs and along mountainsides, the potato's vast and nutritious yields supported the exploding Irish population. Between 1779 and 1841 the Irish population increased by 172 percent, to 8 million inhabitants.[21,27] Ireland had the densest population in Europe, with its arable land having a greater density of people even than that of mid-nineteenth-century China.[27]

A unique vulnerability settled over the land as the peasant class of Ireland, which made up 95 percent of the population, relied almost entirely upon potatoes sown densely in the soil of the overcrowded island.[21] Indeed, the population had grown so much, and the land allotted to subsistence had been divided into such small plots, that potatoes were the only food available that could sustain a poor Irish family. The potato's success led to a vicious, self-perpetuating dependency. "The whole of this structure," wrote Cecil Woodham-Smith about Irish society in a history of the Great Irish Potato Famine, "the minute subdivisions, the closely-packed population existing at the lowest level, the high rents, the frantic competition for land, had been produced by the potato."[27]

One of the first products of globalization, the potato in 1845 became the target of a fungal rot that launched the worst famine ever recorded. The principal food of Irish peasants, seemingly over-night, putrefied into a toxic mush. "History presents no parallel to our circumstances," wrote one Irish resident. "There is no other instance on record of the whole food of a people becoming rotten before it was ripe."[21]

Official British indifference to conditions in Ireland, coupled with centuries of discriminatory policies, created a perfect storm when the potato blight struck in 1845. Discrimination had focused upon the difference of religion between Irish Catholicism and the Protestant Church of England. Irish Catholics had only gained the right to sit in Parliament following the Act of Emancipation in 1829; until then, the Penal Laws of 1695 had "aimed at the destruction of Catholicism in Ireland by a series of ferocious enactments."[27] The Penal Laws barred Catholics from serving in the military, pursuing civic careers, voting, holding political office, purchasing land, even attending school. Catholic-owned estates were "dismembered" by the law's clause that split the land among all sons upon an owner's death, unless the oldest son converted to Protestantism, in which case he would inherit all.

Even following Catholic emancipation in 1829, conditions for most Irish did not improve. Tenants paid their rents with crops of oats, wheat, and barley for their largely absent landlords, while they themselves subsisted almost entirely on potatoes.[21,27] The structure of Irish society discouraged industry and productivity. If a tenant improved the land, those improvements belonged to the landlord, and could serve as a justification for increasing rents, so economic incentive for development was nonexistent. Landlords evicted tenants at will, whether or not tenants had sufficient money for their rents, which embedded in the Irish deep feelings of insecurity and resentment.[27] According to a leading economist of the time, rents owed, or the "hanging gale," maintained the lower classes "in a continual state of anxiety and terror" and acted as "one of the great levers of oppression."[27] Once a family was evicted, their home was destroyed. Families that remained in the ruins were chased out, and when they sought refuge in ditches and holes in the ground, they were chased out again, appearing as little more than vermin in the eyes of British law. The Earl of Clare, a Tory Lord Chancellor, said of the landlords that "confiscation is their common title."[27]

Irish potato crops had partially failed many times, most notably in 1728, 1739, 1740, 1770, 1800, 1807, 1821, 1822, 1830–37, 1839, 1841, and 1844, leading to widespread famine and poor future yields as people ate much of their seed stock.[21,27] But nothing could compare with the potato blight that rolled across the Irish countryside in August and September 1845, after appearing first on the Isle of Wight.[27]

The blight likely originated in Mexico, and was then transported several centuries ago southward to the Andes.[28] The blight likely made its way from South America to the north Atlantic states of the United States in 1841–42, where outbreaks occurred first in the coastal states near Philadelphia and New York in 1843. The blight then crossed the ocean to Europe, either from the United States or South America or both, in 1843–44.

The blight may have arrived first in Belgium with potatoes imported to replace those affected by viral diseases and a dry rot caused by the fungal parasite *Fusarium*, or it may have come first with guano traders, whose field of business began in the 1830s.[28] Regardless, the blight crisscrossed the Atlantic with the speed of the merchant fleet that carried potatoes from one hemisphere to the other.[21,23] Speed may have been a critical element: steam-powered ships had begun regular crossings of the Atlantic Ocean in 1838,[29] just seven years before the Irish famine. And merchants preserved their shipboard stocks of potatoes with ice, further ensuring that the blight could survive the transatlantic trip.[30,31]

The blight may have sailed to Ireland on clipper ships or steamships from Baltimore, Philadelphia, or New York, the epicenter of the North American outbreak. The Irish potato famine then discharged starving and typhus-ridden Irishmen into these same cities almost immediately afterward. But these American societies, unlike that of Ireland, did not depend on the potato for their very existence. It was as if the thirty-two counties of Ireland were held together by one long piece of baling wire, and when the blight snipped that wire, the whole country came apart. Had the famine been engineered, it could not have been better executed. Rather than having policies in place that would have made the famine "transient and partial," governmental actions before and during made it "general and overwhelming."[21] In 1845, approximately one-half of the potato crop failed; in 1846 the failure was nearly total, and people died in droves.

The blight seemed to pass through the country overnight. One priest noted on July 27, 1846, as he traveled across Ireland, that the potato crops "bloomed in all the luxuriance of an abundant harvest."[32] One week later, crossing over the same route, he "beheld with sorrow one wide waste of putrefying vegetation. In many places the wretched people were seated on the fences of their decaying gardens, wringing their hands and wailing bitterly the destruction that had left them foodless." The new harvest, just after the famine of 1845, "melted away in a few short days" into "a stinking mass of corruption."[21,27]

Two successive British governments, one under the Tories and the other under the Whigs, proved unequal to the task of saving Ireland. England's leaders believed that swift and meaningful aid would only worsen the plight of the Irish and the economy of the empire because it would disrupt free trade. The English government also was unwilling to disrupt the power of English absentee landlords to clear the land of their starving tenants. These policies ignited the tinder of Ireland into an inferno. More than 1 million famished, disease-ridden Irish emigrants scattered in all directions.

Many Irish hoped to avoid starvation or "famine fever"—typhus and relapsing fever—by emigrating to England, Scotland, Wales, British North America (Canada), and the United States. Traders who ferried desperate Irishmen abroad often arrived in port with one-quarter, one-half, or even more of their passengers dead from starvation and infectious disease. "The scenes of misery on board of this vessel," wrote an observer of the British ship *Erin Queen*, "could hardly have been surpassed in a crowded and sickly slaver on the

African coast."[21] In debarkation ports, in Liverpool, Glasgow, Quebec City, Montreal, Boston, Philadelphia, and New York, the destitute migrants holed up in the cellars of newly created Irish slums, which became the nexus of typhus outbreaks. Irish immigrants were feared as the source of deadly fever and despised for their squalor.

Irish society could not withstand the combination of famine and disease, and each facilitated the other. Typhus and relapsing fever had always run rampant through Ireland in times of famine. During the potato blight famine, these diseases wiped out entire villages, and the doctors, nurses, and priests who endeavored to help the dying contracted typhus and relapsing fever themselves at alarming rates. Lice transmitted both diseases from person to person, but this fact was unknown at that time. The starving peasants of Ireland, lacking even a change of clothes, lived in dense squalor, conditions in which lice flourished. The famine and its accompanying diseases "suddenly brought the accumulated evils of centuries to a crisis."[27] By the famine's end, well over 1 million Irish people had perished, in addition to the million-plus who had emigrated.[21] "It seemed as if America and the grave," wrote the Irish author of a history of the famine in 1902, "were about to absorb the whole population of this country between them."[21]

The globalization of trade brought the potato and its blight to Ireland. The British developed the tenant system of Irish peasantry, which precluded prudent selection of crops; only the potato could support large families on their small plots of rented land. Unfortunately, without land ownership or secure leases, wealth could not be accumulated across generations. As a final blow, the potato blight left no room for luck. The potato blight instigated famine, and famine invited infectious diseases.

Eighteen forty-five was a hopeless moment in time; neither the potato blight nor infectious diseases related to famine were understood or could be combated with chemicals. With a belief in the spontaneous generation of microbes and plant pathogens, and no inkling of the role of insect vectors of disease, the scientists and doctors of 1845 stumbled toward solutions. Those solutions, it turned out, were not far away, though too late for the Irish; a remarkable shift in scientific thinking lay just around the corner.

Water Mold (1861)

On the first appearance of the blight in the autumn of 1845, Professors Kane, Lindley, and Playfair, were appointed by Sir Robert Peel to inquire in the nature of it, and to suggest the best means of preserving the stock of potatoes from its ravages. The result showed that the mischief lay beyond the knowledge and power of man. Every remedy which science or experience could dictate was had recourse to, but the potato equally melted away under the most opposite modes of treatment.—**Charles Trevelyan**, head of the British Treasury during the famine, January 1848[32]

The blight destroyed all potato varieties in the United Kingdom, where people were frantic to avoid another mass starvation. Since nothing worked to stop the blight, growers devoted themselves to discovering new varieties "which should show sufficient constitutional vigor to hold the disease at bay."[23] William Paterson developed a resistant variety in the late 1850s, the Paterson's Victoria, a "splendid cropper of excellent quality" and "practically immune against the disease." In an 1869 report Paterson wrote, "My own conviction regarding the potato blight is that there is no direct cure for it, but that it is entirely owing to atmospheric action in the plant, and that it will be always more or less subject to it."[23]

Unfortunately, Paterson's varieties and others eventually lost their "pristine vigor" to the blight, and their resistance failed.[23] Besides Paterson's Victoria, there was Nicol's Champion (early 1870s), Sutton's Magnum Bonum (1876), and many other resistant varieties that produced excellent results for a decade or two. Therefore, following the disastrous season of 1879, Lord Cathcart stated that "the production of new varieties is of national importance."[23] Others felt likewise, and many new varieties, such as Findlay's The Bruce,

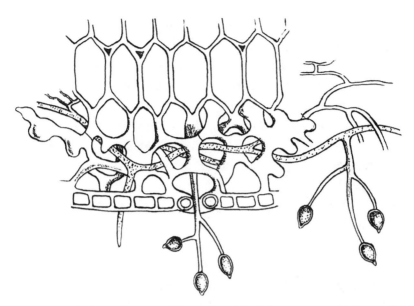

Fig. 1.1. Berkeley's drawing of the potato blight fungus creeping through the underside of a potato leaf.[33]

1836.[36] Berkeley wrote of the potato disease, "It is probable that it has existed for some time without attracting much attention: at any rate it is not the birth of one year only, as the advocates simply of atmospheric influence suppose."[33] Berkeley noted that "the malady is well known in rainy years at Bogota, where the Indians live almost entirely on potatoes . . . The singularity confirms Dr. Morren's notion that the disease, like some other afflictions of the vegetable kingdom, is of American origin."[33]

Berkeley observed in the summer of 1845 that a tiny fungus grew on the diseased plants, and the following winter he declared it the causal agent of blight.[33] His declaration was rejected and ridiculed by most authorities, however, who viewed fungi as a mere result of decay. Berkeley's thesis was therefore tossed aside by critics, one of whom wrote, "The discovery of the cause of the mischief with any certainty seems hopeless . . . and the world has wisely resigned itself to its fate. 'What can't be cured must be endured,' and the potato disease belongs to that class of evils."[35]

Berkeley held his course despite the withering criticisms. "I must candidly confess," he wrote, "that with a becoming share of philosophic doubt where such authorities are ranged upon the opposite side, I believe the fungal theory to be the true one . . . It is by these instruments, contemptible in the sight of man, that the Almighty is pleased sometimes to accomplish his ends."[33] Berkeley received considerable intellectual support from Townley, who advocated restoring vigor to the potato to impart resistance to the fungus.[35] Unfortunately, Berkeley could not demonstrate that the fungus appeared before the blight or how tubers could be infected, because his experimental inoculations were unsuccessful.[33] The theoretical framework for such a claim as Berkeley's required time to mature, and as a consequence his idea remained unresolved for fifteen years.

In fact, long before the Irish famine, Johann Christian Fabricius had already set the framework for the discovery of plant pathogens in his 1774 essay on plant pathology.[37] Fabricius correctly deduced that fungi found in the lesions of diseased plants were separate organisms and not dead plant tissue. Unfortunately, his thesis was rejected by the scientific community for another three generations.[38] By the late 1850s, however, there was general acceptance among scientists of fungi as distinct organisms, and the belief in spontaneous generation collapsed.

The belief in spontaneous generation had persisted for many centuries. Among ancient authors, Aristotle noted that "all dry bodies which become damp, and all damp bodies which are dried, engender animal life."[39] Archelaus wrote that snakes arise from decaying spinal cords.[40] Virgil observed that bees originate from the entrails of bulls. In the seventeenth century, Dutch alchemist Van Helmont wrote, "The smells which rise from the bottom of morasses produce frogs, slugs, leeches, grasses, and other things."[39] All one needed to do to transform wheat into a pot of mice, according to Van Helmont, was to mix a filthy shirt into a vessel of corn; similarly, crushed basil exposed to sunlight could be converted into scorpions. The Italian physician Francesco Redi, however, demonstrated that meat covered with gauze that excluded flies did not generate maggots. This forced the advocates of spontaneous generation to admit that animals that could be seen were not generated spontaneously, but surely, they argued, this could not be the case with microscopic life.

Paradoxically, the invention of the microscope armed advocates for spontaneous generation with a formidable tool. It was common sense that no other process could explain the proliferation of "animalcules"—microscopic animals and other tiny organisms thought to be animals—in decomposing plants and animals.[39] In 1858, Félix Archimède Pouchet, the director of the Museum of Natural History in Rouen, France, announced that he had experimentally proven that microscopic life arose spontaneously upon hay heated to boiling in a sealed vessel of water and oxygen that he had inverted into a basin of mercury. The mercury prevented outside air from entering and therefore contaminating the vessel of hay, and yet microbes emerged. Pouchet had considerable influence, but his proof would soon be crushed.

The idea of spontaneous generation stifled progress on many fronts, including investigation of the potato blight. Berkeley recognized this and noted in 1846 that for his thesis to be incorrect, one need take "recourse to the notions entertained by many of spontaneous or equivocal generation from languid or diseased tissues; for the question at last reduces itself to this, which is indeed one involved in mystery, but which, as far as I can judge, wherever the veil is partially lifted up, seems after all to point to the same general laws by which the higher portions of the creation are governed."[33]

In 1859, Charles Darwin lifted that veil with his landmark book, *On the Origin of Species by Means of Natural Selection, or the Preservation of Favoured Races in the Struggle for Life*.[41] Darwin provided the theoretical framework for the idea that all life was connected by the process of evolution, which would not be the case if new life forms continuously arose by spontaneous generation. Also in 1859, Louis Pasteur jumped into the controversy over spontaneous generation. Pasteur's confidant, Jean-Baptiste Biot, upon hearing from Pasteur that he planned to study spontaneous generation, argued vigorously against such a waste of time. "You will never escape from it," Biot prophesied.[39]

But Pasteur did; he performed experiments in which he used a glass bulb with a long, bent neck attached to a heated platinum tube.[39] Air could only enter the glass bulb after first passing through the tube, which due to its extreme heat killed all germs. Pasteur showed that when germs were thus excluded from a nutrient broth that had first been sterilized in the heated glass bulb, the broth remained sterile; germs did not arise spontaneously. He tried this experiment with different "putrescible liquids," including urine. He then extracted dust from the air and showed that sterilized nutrient broths could be seeded with this dust and the germs it contained to become fertile. He repeated his experiments under varying circumstances that consistently showed that microorganisms had to colonize the nutrient broth in order for it to produce life. For his experiments disproving spontaneous generation, Pasteur won the 1860 prize from the French Academy of Sciences, which was offered to the person who could best "endeavor by well-contrived experiments to throw new light upon the question of spontaneous generation."[39]

Even these experiments, however, did not convince everyone; if the smallest bubble of air held germs, Pouchet argued, then germs in air "would form a thick fog, as dense as iron."[39] To this objection, Pasteur performed experiments where a nutrient broth was exposed to air from different localities, some of which, like those at high elevation, rarely contained germs. In a keynote lecture given in 1864, Pasteur criticized Pouchet's experimental work.[39] "This experiment is irreproachable," he said, "but irreproachable only on those points which have attracted the attention of its author." Pouchet had made an error that "renders his experiment as completely illusory as that of Van Helmont's pot of dirty linen. I will show you where the mice got in." The microorganisms got into Pouchet's flask, Pasteur showed, on particles of dust attached to the mercury.

After defeating all possible arguments for spontaneous generation, Pasteur announced, "Those who maintain the contrary have been the dupes of illusions and of ill-conducted experiments, tainted with errors which they know not how either to perceive or to avoid. Spontaneous generation is a chimera."[39] Pasteur's work on spontaneous generation had profound practical applications, including his demonstration that microorganisms could be killed with heat, a technique that came to be known as pasteurization.[42]

Pasteur devoted himself to finding cures for diseases following the deaths of three of his five children, two of whom succumbed to typhoid fever.[42] Pasteur reasoned that just as microorganisms spoil milk, so, too, could they be responsible for the onset of disease. In the 1870s, Pasteur and his intellectual rival in Germany, Robert Koch, independently demonstrated that a microorganism causes anthrax, providing proof of the germ theory of infectious disease.[43] The Pasteur-Koch rivalry was more than personal; the two men represented the prestige of their nations in fiery academic debates immediately after the defeat of France by Germany in the Franco-Prussian War. Koch won the Nobel Prize in 1905. Pasteur died before the first Nobel Prize was awarded, or he, too, surely would have received one.

The theory of Pasteur and Koch that germs cause infectious disease had an immediate and critical corollary: that it must be important to prevent microorganisms from gaining entry into the body. This logical deduction led Joseph Lister to his surgical advance of antiseptic techniques, which he published in 1867.[44] Meanwhile, Pasteur developed vaccines for anthrax, chicken cholera, and rabies, the last of which he first tried in 1885, with success, on a nine-year-old boy named Joseph Meister who had been bitten by a rabid dog.[42] The Pasteur Institute subsequently employed Meister as gatekeeper; fifty-five years after Pasteur saved his life, Meister committed suicide so that he could not be forced to open Pasteur's crypt for Nazi invaders.

It was hard for people to imagine that infinitesimally small life forms could wipe out the crop of a nation or cause epidemic disease. Pasteur broke through that intellectual barrier by showing that germs do not arise spontaneously and that they are the basis of human disease. Surely the same principles could hold true for plant diseases.

The same year that Pasteur published his landmark experiments disproving spontaneous generation, another remarkable microbiologist discovered the pathogen that blighted Irish potatoes.[38] Anton De Bary first gained prominence in 1853, when, just twenty-two years old, he published his classic work demonstrating that fungi associated with rust and smut plant diseases were not generated spontaneously and were the cause and not the effect of disease. Over the course of his career, De Bary established the causal link between particular fungi and a remarkable array of plant diseases. De Bary's work established the modern field of mycology (the study of fungi), and was a precursor to Pasteur's work on spontaneous generation. In 1861, De Bary published his study showing that *Phytophthora infestans* was the pathogen responsible for potato blight.[36,45]

The blight pathogen was first described and named by an old surgeon from Napoleon's army, Dr. Montagne, at the Société Philomathique in Paris on August 30, 1845.[36] He called it *Botrytis infestans*, and because he beat the competition by a few days, the other proposed names, *Botrytis vastatrix* and *Botrytis fallax*, lost out. Montagne's description of the pathogen was published in Berkeley's key treatise on the subject.[33] The name was later changed to *Peronospora infestans*.[36] De Bary discovered key differences between the blight pathogen and other members of the *Peronospora* genus that warranted its membership in a separate genus. He renamed the blight pathogen *Phytophthora infestans*, imparting a new name to the species after its behavior: *phyto* means "plant," *phthora* means "corrupter," and *infestans* means "aggressive," "hostile," or "dangerous."

Phytophthora infestans is not truly a fungus, but is a fungus-like organism belonging to the group Oömycota, or the water molds.[46] De Bary inoculated potato leaves, stalks, and tubers with spores of the pathogen and then followed its progress through to potato rot.[36] De Bary demonstrated that *Phytophthora infestans* enters the potato leaf through the small holes that allow gas exchange in plants (individually known as "stoma," or in the plural "stomata," derived from the Greek word for "mouths"). The parasite's mycelium (the vegetative part composed of threadlike hyphae) then branches, its numerous hyphae worming their way among the leaf's cells and consuming their nutrients.[23] Hyphae then extend outward through the stomata and produce branches tipped with pear-shaped fruiting bodies (sporangia). The sporangia easily detach and fall onto the soil or are carried aloft by the wind. When they land on another potato leaf, they bide their time until a drop of water from rain or dew stimulates their growth. The water mold then holds two options. In the first, it may grow a filament that enters the leaf through a stoma, repeating the process of its own formation. In the second, it may develop and release infectious spores. These ride droplets of water through soil particles until they meet the potatoes below and initiate rot, which is why many people attributed the blight to excessive dampness.

The fact of infection does not become evident to the farmer until yellow or brown spots form on the leaves, "much as if a dilution of acid had fallen upon them like drops of rain."[21] The spots enlarge and blacken before the leaf curls, decomposes, and produces a unique and unpleasant smell. Words describing the terrible smell typically found their way into the alarming reports of Irish potato failure. The mildew that borders the rotten leaf tissue is the forked hyphae tipped with sporangia ready to accelerate the field's infection until it appears as if a fire has swept across it. Berkeley noted that the blight pioneers the potato's destruction, for following infection, "other fungi establish themselves on the surface, or in the decaying mass, which emits a highly fœtid odour resembling that of decaying agarics, the union of the cells is dissolved, animalcules or mites make their appearance, till at last the whole becomes a loathsome mass of putrescence."[33]

The Irish summer of 1848 mirrored those of 1845 and 1846. "On the morning of the 13th [of July]," wrote a parish priest, "to the astonishment of everyone, the potato fields that had, on the previous evening, presented an appearance that was calculated to gladden the hearts of the most indifferent, appeared blasted, withered, blackened and, as it were, sprinkled with vitriol, and the whole country has in consequence been thrown into dismay and confusion."[27]

Reports came in of potato fields "blackened as if steeped in tar," and as before, the scarred landscape carried an "intolerable stench."[27] Potatoes "were either positively rotten, partially decayed and swarming with worms, or spotted with brownish coloured patches, resembling flesh that had been frost-bitten."[33]

In the late nineteenth century, many evolutionists incorrectly argued that things in nature have a purpose for which they evolved, even very bad things. It was a philosophy that allowed a hand to guide the evolution of life, a marriage of a faith in Creation with the modern tenets of evolution. It gave people comfort to think that evolution progressed with purpose, and the philosophy was consistent with humankind occupying a position at the pinnacle of Creation. This philosophy also allowed people to view potato blight as a positive force that killed sickly varieties and furthered the propagation of robust varieties. At the time, fungi were believed to be plants. Nine years after De Bary's discovery of *Phytophthora infestans*, a potato cultivation expert wrote, "The object of nature in bringing into existence the large family of mildews, each member of which is a perfect plant in its way, and as capable of performing its functions as the oak of the forest, was undoubtedly to prevent propagation from sickly stock, and by the decomposition of feeble plants to make room and enrich the soil for the better development of healthier plants . . . The potato disease is rather an effect than a cause, and appears to have been designed to prevent members enfeebled by accident or otherwise from propagating their species by putting such members out of existence."[47]

Although such ideas of evolutionary purpose and progress persisted well into the twentieth century, they were a misinterpretation of evolutionary theory as presented by Darwin. Indeed, Pasteur, Koch, De Bary and Darwin broke through long-established modes of thinking stifled by the concepts of special Creation and fixity of forms. These mid-century paradigm shifts—that life did not arise spontaneously from inanimate matter, that microscopic organisms could be the source of disease of both plants and animals, and that life evolved—fed a furious pace of scientific discovery that has never ceased.

Bordeaux Mixture (1883)

Since the appearance of mildew in France in 1878, I have not ceased to study *Peronospora* in the hope of discovering in its development some vulnerable point that might permit of its mastery.—**Pierre Marie Alexis Millardet**, 1885[48]

De Bary's identification of the pathogen responsible for potato blight brought to light a tempting target. If an agent could be developed to kill *Phytophthora infestans*, future potato famines could be averted. The development of such a weapon followed a curious path that had nothing to do with potatoes.

A tiny relative of the aphid, called *Phylloxera* (from the Greek for "the drying up of leaves"), made its living by sucking sap from the leaves and roots of grape plants in its native lands of eastern North America.[36] When English wine enthusiasts imported botanical specimens of American grapevines in the 1850s, *Phylloxera* hitched a ride. But unlike their American counterparts, European vines were not resistant to the insect. In 1865, vines in France withered before the onslaught of *Phylloxera* as the Great French Wine Blight cast wineries into ruin. In the years that followed, wine production tumbled across Europe, ultimately affecting about 2.5 million acres of vines. French vineyards employed numerous techniques to stop the destruction, including burying live toads under the vines to pull out the plants' poison, and injecting carbon disulphide into the soil, but to no avail. Nothing, it seemed, could stop *Phylloxera*—also known as plant lice.

Pierre Marie Alexis Millardet, a French botanist and mycologist, confronted the problem of the wine blight. Millardet was perfectly suited to the task; he had studied medicine in Paris—a prudent choice to support his mother and younger siblings after his father died in the 1854 cholera epidemic.[49] At that time, medical school necessitated the study of medicinal plants. As a young physician, Millardet studied in Heidelberg, and then worked in Freiburg for De Bary, before receiving an appointment in 1869 as a professor of botany at the University of Strasbourg.[50,51] The following year, with the outbreak of the Franco-Prussian War, Millardet enlisted as a medical officer in the French army. France suffered a humiliating defeat and lost Alsace and much of Lorraine; furthermore, Strasbourg was now a German university.[51] Millardet therefore accepted a position in Nancy in 1872, and in 1874 he traveled to Bordeaux to investigate the *Phylloxera* infestation. Two years later, engaged in the study of *Phylloxera*, Millardet relocated to the University of Bordeaux for the position of chair of botany.

Millardet demonstrated that by grafting American grapevines onto French vines, hybrids resistant to *Phylloxera* were created.[50] Unfortunately, in 1878, Millardet discovered that some American vines imported to France were infected by another American import, the downy mildew, which, like *Phytophthora infestans*, is a water mold. Downy mildew concentrated its pathogenic efforts where *Phylloxera* hesitated, and sped across France in a path of vineyard destruction. Millardet's focus shifted from conquering *Phylloxera* to defeating downy mildew.

In October 1882 Millardet observed that vines along roadsides of the vineyard of Saint-Julien-en-Médoc did not exhibit mildew on their tissues, while rot initiated by downy mildew remained chronic away from roads.[48] Millardet questioned the viticulturists about the "pulverized bluish white substance" that he observed upon the leaves of the uninfected plants.[48] They told Millardet that they sprayed a visible and bitter-tasting mixture of lime and copper sulfate along the roads to discourage foot-travelers from pilfering their grapes.[51] It was a stroke of great luck that this mixture also protected the plants from downy mildew.

Millardet, armed with this serendipitous discovery, conducted experimental trials with various combinations of the different salts of copper and iron with lime powdered or dissolved.[52] He demonstrated that a particular mixture of copper sulfate and lime, which came to be known as Bordeaux Mixture, was an effective solution against downy mildew, and did not harm the plant or grapes.[51] Millardet worked with the chemist Ulysse Gayon to determine the perfect concentrations of ingredients for maximum effectiveness.[51,52]

The downy mildew, upon settling on a grape leaf treated with Bordeaux Mixture, either dies directly or its zoospores are unable to generate their germ tubes and penetrate the epidermis of the grape leaf.[52] Infection is averted. "The leaves are healthy and of a beautiful green, the grapes are black and perfectly ripe," wrote Millardet. "The vines that were not treated present, on the contrary, the most wretched appearance, the majority of the leaves have fallen; the few that remain are half dried up; the grapes, still red, will not be fit for anything except sour wine."[48]

Millardet also found early in his investigations that the reproductive bodies of downy mildew did not develop in water from his household well, though they did in city water, rain water, dew, and distilled water.[52] The explanation for this became apparent only after his discovery of Bordeaux Mixture. Water was drawn from his well by means of a copper pump, and the water, which contained 5 mg of copper per liter, also held dissolved limestone from the surrounding rock.[52] Millardet's well, by happenstance, produced its own Bordeaux Mixture because it contained both copper and lime.

Bordeaux Mixture was cheap and effective; 50 liters was sufficient to treat 1,000 plants at a cost of only 5 francs for materials and labor, and "one need fear no harm, even to the most tender organs" of the vine.[48] A single treatment was all that was needed to protect vines from downy mildew, but it had to be applied preventively since the mildew grows within the grape leaves.[48,52] Millardet's work gave rise to the world's first commercial fungicide, and more generally, to the first pesticide effective against plant pathogens.[50,51]

Millardet was most concerned about the priority of his findings over those of Baron Chatry de la Fosse and others, who also observed, but two years later, that grapes treated with this mixture did not succumb to downy mildew.[53] Just as Darwin accelerated the publication of On the Origin of Species because Alfred Russel Wallace had independently discovered natural selection and threatened to scoop him, so, too, did Millardet rush his work to publication.[48] Millardet felt it necessary to write a detailed chronology of who discovered what and when, month by month, in order to establish his precedence in the discovery of Bordeaux Mixture.[53] "I claim the honor," he wrote, "of having conceived the treatment with copper, that of having first experimented and, likewise, of having first proposed the practice. May I be permitted to add—for these are for us learned men our titles and our dearest souvenirs—that first, in 1878, simultaneously with M. Planchon, I observed the presence of mildew in France. Since then I have constantly been on guard."[53]

In the small world of nineteenth-century public health research, considerable cross-talk linked investigations between scientists and nations. Some advances were ignored at tremendous cost to human lives. Others were seized upon and broadly applied. The latter was the case with Bordeaux Mixture, which sped in its application from the grapes of France to the potatoes of Ireland. Indeed, Millardet anticipated this critical outcome. "The close analogies that exist between the *Peronospora* of the vines," he wrote, "and that causing the disease of the potato and tomato cause me to hope that we shall henceforth have at hand a real prophylactic treatment for these latter diseases."[48] What worked on grapes worked on potatoes. Researchers in Ireland snatched up the idea and sprayed potato crops with Bordeaux Mixture to combat *Phytophthora infestans*, which "proved up to the hilt that this system was of incalculable value in either altogether preventing or at least very materially checking the ravages of the disease."[23]

Bordeaux Mixture was simple to prepare: 12 pounds of copper sulfate was neutralized with approximately 8 pounds of freshly burned quicklime, both previously dissolved in a total of 75–100 gallons of water. With Bordeaux Mixture applied, farmers harvested usable crops even during years of heavy blight, as long as they sprayed the mixture before the blight manifested itself and at intervals throughout the season. Had Bordeaux Mixture been available in 1845–49, Ireland's misery could have been averted, America would be a different place without an immense immigrant Irish population, typhus epidemics would not have torn through cities where Irish immigrants alighted, and the bitter divide between Ireland and England would have been less acrimonious. Indeed, some scientists had tried applying salts of copper upon potatoes during the famine,[35] but not in the right concentrations or mixtures.

Phytophthora infestans, or the late blight, as it turns out, was not the only enemy of the potato faced by nineteenth-century farmers. Various other diseases, such as early blight, potato scab, dry rot (*Fusarium* blight), stem rot (rhizoctonia), black leg, and European wart disease, infested potato fields.[23] Insect enemies were also plentiful and included potato flea beetles, Colorado potato beetles, potato bugs, potato worms, potato stalk weevils, potato eelworms, grasshoppers, and even a species, the three-lined leaf-beetle, whose larvae cover themselves with their own excrement.[47] In some cases, insect larvae forged the paths into potatoes that were followed by the pathogenic fungi. Just as *Phytophthora infestans* followed the potato around the world, so, too, did many of these insect pests. "When civilization marched up the Rocky Mountains," wrote a potato expert in 1870, "and potatoes began to be grown in that region, [the Colorado potato-bug] acquired the habit of feeding upon the cultivated potato. It went from potato-patch to potato-patch, moving east-ward at the rate of about sixty miles a year, and is now firmly established over all the country extending from Indiana to its old feeding-grounds in the Rocky Mountains. In about twelve years it will have reached the Atlantic coast."[47]

Although the potato farmer faced many enemies, Bordeaux Mixture was not the sole chemical tool in the turn-of-the-century gardening shed. Other pesticides were also available, such as bichloride of mercury, formaldehyde, Paris green, arsenate of lead, and arsenic-bran mash.[23] And Bordeaux Mixture proved useful beyond preventing late blight. It was also applied against other potato pathogens, and it induced vigorous growth. "No matter where the crop is grown," wrote one expert, "or whether diseases are present or not, the writer feels warranted in recommending the application of the mixture, on the ground that its use will yield a handsome return."[23] Another expert recommended the use of Bordeaux Mixture "freely and vigorously," with the addition of arsenate of lead to kill potato bugs and larvae.[23] The combined mixture should be sprayed from all directions, for "flank an enemy and he is in a dangerous condition."

Advances in the chemical prevention and treatment of potato diseases, along with coincident advances in farming technology, such as the Iron Age Traction Sprayer, Riding Cultivator, Potato Planter, and Potato Digger, increased yields throughout North America and Europe.[23] The technology of mechanized pesticide applicators advanced along with the Industrial Revolution so that there always seemed to be new devices to buy. Increased yields from spraying could not be ignored. For example, American scholars calculated in 1912 that a potato field on the island of Jersey sprayed with Bordeaux Mixture five times during the growing season at a cost of $1.25 each time yielded 13 tons to the acre, while an adjacent field only sprayed twice was scorched with blight.

Enthusiastic growers sought perfection in "a spud that nicely fills the fist of a good-sized man."[23] "Next to bread and meat," according to a popular booklet from Colorado, "the most important article of food to the Anglo-Saxon race is the potato."[23] Given that demand, efficient production methods utilizing pesticides yielded excellent profits. Demand for potatoes grown in Washington state at the turn of the century kept their price above $10 per ton, and farmers were able to secure net profits of $15–$20 per acre.[23] The focus shifted from famine in 1845 to profits half a century later because of the sweeping agricultural revolution brought forth by pesticides.

The Irish Potato Famine of 1845–49 impelled the search for the responsible pathogen. Pasteur disproved spontaneous generation in 1861, and that same year De Bary discovered that the water mold *Phytophthora infestans* caused the potato rot. Two decades later, Millardet developed the world's first effective fungicide. Bordeaux Mixture first saved European vineyards, then averted additional potato famines, and finally boosted profits in the global business of potato production. Millardet proved that people could make chemical agents that were almost magical in their application. The door to fighting famine and infectious disease was thrown open, and a host of brilliant scientists charged through it, determined to conquer humanity's most pressing problems.

PART 2

Plague

[**2**]

Marsh Fever
(2700 BCE–1902)

Viewed in the light of our modern "germ theory" of disease, the punctures of proboscidian insects, like those of Pasteur's needles, deserve consideration, as probable means by which bacteria and other germs may be inoculated into human bodies, so as to infect the blood and give rise to specific fevers.—**Albert Freeman Africanus King**, 1883[54]

Malaria co-evolved with our human ancestors in Africa and accompanied anatomically modern humans as they migrated to Eurasia and later to the rest of the world.[55] Malaria was likely the first and most deadly infectious disease to sweep through ancient human settlements that formed with the advent of agriculture.[56] Reliance on water sources put villages and cities adjacent to the breeding grounds of the *Anopheles* mosquitoes that carry malaria. Further, the population density facilitated by agriculture allowed accelerated infection rates.

The progression of malaria through its stereotypical periodic fevers is described as early as 2700 BCE in Chinese medical texts.[55] The Greek physician Hippocrates described malaria in detail in the fifth century BCE; he considered many variables when judging a disease state, paying attention to "the whole constitution of the season, and particularly to the state of the heavens," the patient's dreams, and to "flatulence, whether passed silently or with a noise."[57] Other distinguished authors of the ancient world, from India, Assyria, Arabia, Greece, and the Roman empire, advanced knowledge of the disease and noted its association with marshes.[55] Malaria came to be known as "marsh fever."

The marshland association led to many hypotheses to explain malaria transmission, most of which involved "miasmas," the term created by Hippocrates for poisonous fumes rising from the earth.[55] The belief in poisonous air as the cause of malaria led to its name, which is derived from the Italian *mala aria*, or "bad air."[58] Miasmas as the cause of malaria appeared often in literature, as in Shakespeare's *The Tempest*: "All the infections that the sun sucks up / From bogs, fens, flats, on Prosper fall, and make him / By inch-meal a disease!"[59]

"Cures" for malaria were diverse and often magical. Quintus Serenus Sammonicus, the physician to the third-century CE Roman emperor Caracalla, advised sufferers of malaria to wear an amulet with the inscription "abracadabra" for nine days, then to toss it over their shoulder into a stream running in an eastward direction.[60] Following this, lion's fat could be applied to the patient's skin, or the skin of a cat, adorned with yellow coral and green emeralds, could be placed about the neck.

Although ancient people did not understand the role of mosquitoes in malaria transmission, they observed disease patterns and altered their behavior accordingly. The Romans attempted to drain wetlands, and the Abbasid caliphate successfully reduced malaria in Baghdad through drainage.[56] In Southeast Asia, people constructed their homes on stilts, which placed them above mosquito flight paths. Malaria also structured cultural landscapes. Its association with wetlands and lowlands led many mountain people to stay at elevation during the malaria season, thereby strengthening the cultural divides between people of the mountains and people of the lowlands.

The African slave trade conveyed the deadly falciparum variety of malaria to the Americas, where it joined other Old World diseases in wiping out most of the indigenous population.[56] African slaves had genetic resistance to malaria and acquired immunity to a host of tropical diseases, so the population crash brought about by the diseases they introduced to the Americas accelerated the slave trade. African slaves largely replaced both indigenous slaves and European indentured servants, who had no immunity; the disease resistance of Africans ensured their downfall into slavery. And because Africans had both genetic and acquired resistance to tropical diseases, they were regularly employed to crew slave ships.

Malaria worked in concert with other diseases to set the course of political events in the American colonies. British forces overwhelmed the Spanish on Jamaica in 1655, but by the following year most of the British soldiers had fallen to malaria and dysentery.[56] By the turn of the seventeenth century, malaria killed more people in the North American colonies than any other disease.[61] When the British invaded the French colony of Saint Domingue in 1794–95 to put down a slave revolt, they lost one hundred thousand soldiers to malaria and yellow fever; the British were so weakened that in 1801 the former slaves founded the nation of Haiti.[56] When the French then invaded Haiti in 1802, malaria and yellow fever whittled down Napoleon's force of sixty thousand soldiers to fewer than ten thousand.

Malaria played similar roles in a long list of wars, including wars between kingdoms in Europe, colonial wars in Africa, the American Civil War, World War I, the Russian Civil War, and World War II.[56] A historian of the effects of malaria on warfare wrote in 1910 of the 1864 British campaigns in West Africa, "It can scarcely be called a war, as an enemy was never seen, or a grain of powder expended; our troops were defeated by disease, much of which was preventable."[62] In Madagascar in 1895, the French suffered thirteen deaths in battle and more than four thousand deaths from malaria.[63] During the battles over Macedonia in World War I, malaria paralyzed the French, British, and German armies for three years; almost 80 percent of the French soldiers were hospitalized with the disease, while the British had 162,512 hospital admissions for malaria compared to a total of 23,762 combat casualties or solders taken prisoner or missing. Following each war, infected soldiers trudged home and were bitten by mosquitoes that then initiated new malaria outbreaks.

Even the political landscape that inevitably led to war could be blamed to some degree on malaria. For example, heightened malaria rates in the southern US states accentuated the economic drive for slavery since Africans were resistant to the disease; more Africans brought more malaria and created an epidemiological divide whose geography mirrored the political divide between North and South.

Malaria could not be separated clinically from other diseases that caused similar symptoms. This posed a formidable barrier to malaria research that was finally overcome in the early seventeenth century when Jesuit monks learned from the indigenous people of Peru that the bark of a South American tree effectively treated certain types of fevers known as "intermittent fevers"—that is, malaria. The tree was called the cinchona tree after the name of the Peruvian viceroy's wife, who was cured with its bark. The Jesuits imported the bark to Europe around 1640, but their treatment was ridiculed by most men of science.[64] Nonetheless, this discovery allowed astute investigators in the early eighteenth century to separate malaria from other sources of fever and thereby to chart the course of the illness more precisely than had been done before.[65,66]

A British medical officer during the American Revolution wrote of the controversy regarding cinchona bark, "The Hessians were all of them inveterate enemies to the bark; and there were ever some of the British surgeons who employed it very sparingly . . . There was a Hessian regiment . . . that lost one third of its men by this disease and its effects, during one year's service in Georgia. There were British regiments also, which lost more than a fourth; while there were others, which did not lose a twentieth. The whole of these regiments were engaged on the same services; they were all alike foreigners in America; and there appeared to be no obvious case for so great a difference in the degree of mortality, except a difference in the management of the bark."[67]

In 1820 the French chemists Pierre-Joseph Pelletier and Joseph Bienaimé Caventou extracted two of the four active ingredients of cinchona bark, quinine and cinchonine.[56,64,68] Soon afterward, cinchona plantations and quinine extraction became big business. Chemical firms sprang up in Europe and the United States to produce quinine, and the modern chemical industry was born. Quinine became the first Western medicine produced to treat a specific disease and so paved the way for the modern pharmaceutical industry.[56] The widespread availability of quinine also facilitated European colonization of Africa and US conquest of Native American tribes.

The actual agent of malaria could not be discovered until after the invention of the microscope in the late seventeenth century, and even then, it took two centuries longer to make the discovery. First, in the 1850s, with the aid of microscopes, researchers discovered that malarial patients harbored black-pigmented particles in their blood, which they called melanin.[69] Pasteur and Koch then set the theoretical stage in the 1870s; the development of the germ theory of disease led to the rapid discovery of pathogenic bacteria that caused a host of diseases, including anthrax, relapsing fever, tuberculosis, pneumonia, typhoid fever, diphtheria, tetanus, and Asiatic cholera—all by 1890. Bacteriologists also chased down the microbe that causes malaria, but the investigation into bacteria as the culprit turned out to be a false start since the microbe is not a bacterium.

The critical clue to the malaria pathogen was the presence of melanin in the blood of malarial victims. The person who followed the melanin trail, the French army surgeon Charles Louis Alphonse Laveran, started his stellar career in 1870, when, at the age of twenty-five, he served the French military as a medical assistant major in the Franco-Prussian War.[69] Following the war, Laveran obtained the position of chair of military diseases and epidemics at a military medical school, which had previously been held by his father. In 1878, he was dispatched to a French military hospital in Algeria, where he set out to determine if melanin was only found in malarial patients and hence could be used as a diagnostic tool. His close inspection of the blood of malarial patients led him in 1880 to discover a previously unknown entity, which he determined was the malaria parasite.[70] Even though Laveran lacked staining chemicals to help visualize the parasite, he found that it developed in the red blood cells and destroyed them, which altered their pigmentation from red to black and thereby produced the melanin.[69]

Laveran found that the malaria parasite was not a bacterium, but rather belonged to the protozoa. In 1882, Laveran moved to a marshland of Italy, and there again found the same parasite in the blood of malarial patients, and so gained certainty that this was the agent of malaria. He published this conclusion in 1884.[71] Although other scientists were at first skeptical, they were able to replicate his research, and in 1889, with his conclusion now accepted as fact, the French Academy of Sciences awarded him its Bréant Prize.[69]

Laveran's timing was perfect, and he initiated a new branch of infectious disease research focused on the protozoa. This led to the discovery that a group of protozoa known as trypanosomes (named for the Greek "body borer") cause numerous animal and human diseases, the most important of which is sleeping sickness.[69] Because various species of flies are the vectors of trypanosome diseases, his discoveries also provided important targets for pesticides. Laveran received the Nobel Prize in 1907; he donated half of his prize money to establish the Laboratory of Tropical Medicine at the Pasteur Institute.

Laveran knew that his identification in 1880 of the malaria pathogen in humans had only begun the chase for a cure. His next objective was to find the parasite outside the human body. He searched in vain in the water, soil, and air of marshlands.[69] Other scientists tried inoculating human subjects with filthy marsh water to infect them with malaria.[65] The absence of the parasite in the marshlands led Laveran to conclude in 1884 that the parasite must be carried by mosquitoes.[69] After all, some agent was necessary to move the parasites from the blood of one person to that of another, and mosquitoes were ubiquitous in marshes. Other prominent scientists converged on this idea at about the same time.

Patrick Manson, the Scottish physician considered to be the father of tropical medicine, determined in 1876 in China that mosquitoes transmit *Filaria* worms, which cause lymphatic filariasis, also known as elephantiasis.[72] Manson's gardener was infected, so Manson had mosquitoes feed on him while he slept; he found the parasite in both the gardener and the mosquito and described the *Filaria* life cycle. Amazingly, the *Filaria* embryos hid during most of the day in blood vessels deep inside the body, then appeared in the peripheral blood in immense numbers from just after sunset until about midnight. "It is marvelous," wrote Manson, "how Nature has adapted the habits of the filariae to those of the mosquito. The embryos are in the blood just at the time the mosquito selects for feeding."[73] Based on his work with *Filaria*, Manson hypothesized in 1894 that mosquitoes were also the culprit in malaria transmission.[74] Living in England at the time, he could not test the idea himself, so he wrote, "I would commend my hypothesis to the attention of medical men in India and elsewhere, where malarial patients and sectorial insects abound."[74]

Koch also thought mosquitoes served as vectors for malaria, but he did not publish his hypothesis.[65] In fact, the suggestion had been made much earlier, in the early nineteenth century.[54] Some records even show the idea of insects as vectors for malaria being advanced more than two thousand years ago, and Italian peasants had blamed the mosquito for centuries.[58] Koch found the same belief prominent among natives of the central African highlands, and in Ethiopia the native elephant hunters reported that they could safely pass through malarial regions because they daily fumigated themselves with sulfur.[75] Similarly, sulfur miners in Sicily suffered a small fraction of the malaria infections experienced by residents engaged in other occupations, and a Greek city of forty thousand people collapsed from malaria deaths after a local sulfur mine closed.

The person who best developed the mosquito hypothesis was Albert Freeman Africanus King (so named because of his father's fascination with colonization of Africa). King was a twenty-four-year-old assistant surgeon in the US Army in attendance at Ford's

Theatre in Washington, DC, when John Wilkes Booth shot Abraham Lincoln; King was one of three physicians who attempted to treat the mortally wounded president.[76]

King independently developed the mosquito hypothesis in 1883 based on the tight association between the geography of malaria and the life history of the mosquito.[54] Most of King's analyses were spot on. King pointed out that "germs are so minute that a million may rest on the head of a pin."[54]

Although King's development of the mosquito hypothesis was the most thorough of his time, some of his analyses were based on fanciful nineteenth-century ideas of nature's purpose. He argued, for example, that poisonous air could not be the source of malaria for "it was surely never designed that breathing the indigenous air of any natural environment should be, and without warning of danger, a means of death."[54] Nature provided warnings of danger, such as the rattle of a snake. "Man," he wrote, "naturally loves beautiful things, woman and the flowers. But the serpent also is beautiful, superficially smooth, tapering in form, elegantly elastic, absolute in symmetry, undulating in motion, every element of beauty in woman finds its counterpart in the snake; yet we love the one, loath the other."

Based on his exhaustive examination of the mosquito's probable role in malaria transmission, King proposed a series of protective measures, including the destruction of mosquitoes. But King was not an experimentalist and did not have the means to test his ideas.[77] His ideas were disregarded.

Anopheles (1894–1902)

> The world requires at least ten years to understand a new idea, however important or simple it may be. The mosquito theorem of malaria was at first ridiculed, and its application to the saving of human life treated with neglect, jealousy and opposition.—**Ronald Ross**, 1910[78]

The experimental development of the mosquito hypothesis fell upon an unlikely Scottish scientist named Ronald Ross. Ross, the son of a British general, began working in the Indian Medical Service in 1881 after a mediocre academic performance.[69,79] Although Laveran had already identified the malaria parasite a year earlier, Ross did not have access to cutting-edge science in his remote Indian outpost and so he was unaware that a parasite was involved.[65] Ross's observations in 1889 that the geographic patterns of malaria infection were inconsistent with the old poisonous-air hypothesis motivated him to study the disease in detail. First, he published papers outlining his hypothesis of "intestinal auto-intoxication."[80–84] Then, in 1892, he became aware of Laveran's discovery of the parasite, though he did not believe it until 1894 when he returned to England and was convinced by a mentor that Laveran was correct; Ross's mentor told him to get in touch with Manson.[65]

Manson showed Ross the parasite and told him of his mosquito hypothesis.[65] Ross knew that the problem of identifying the mode of malaria transmission was paramount, and he found himself "immediately and powerfully struck" with the mosquito hypothesis.[65] He recalled and related to Manson that Laveran had the same idea. The mosquito hypothesis, Ross noted, was even "held by some barbarous tribes who have only too great an experience of the disease."[85] "Ignorant of the route of entry," wrote Ross, "we could rest our prophylaxis only upon an unsatisfactory empirical basis; cognizant of it, we might hope to stamp out the plague even in its most redoubtable haunts."[65]

Ross and Manson hatched a plan. Ross would return to India, allow various species of mosquitoes (for they did not know which might be to blame) to feed upon malarial patients, and track the parasite in the mosquito's tissues and into the water where the mosquito deposited her eggs.[65] From there, Ross could determine how the parasite moved from water into human beings. This plan was based on Manson's erroneous belief that mosquitoes only took one blood meal before depositing eggs and dying, and therefore they could only communicate the parasite from humans to water and subsequent infection must be through ingestion of water.[74]

Ross returned to India in 1895 to serve as the medical officer of a regiment of Indian soldiers who harbored high rates of malaria infection.[65] British India in the late nineteenth century experienced dramatic drought-induced famine (1876–78 and 1896–1900).[56] In a reflection of its treatment of Ireland, the British government allowed "market forces" to contend with the problem, and somewhere between 12 million and 29 million Indians perished. As with Ireland, the agent of death was usually fever, this time due to malaria.

Ross raised mosquitoes in captivity to be sure they were not previously infected, then let them feed on malarial patients. He dissected the mosquitoes at various time points after feeding in order to create a time series of the parasite's development within the mosquito.[65] By the end of 1895, after excruciating work over his microscope, Ross found himself unable to locate the malaria parasite in mosquito tissue.[86] Frustrated, Ross realized that he must not have any preconceived notions of what to look for—perhaps the parasite in the mosquito looked nothing like its form in people. "The protean changes of many of the parasitic worms," he wrote, "warned us that nature was capable of ordering any extraordinary transformations in the interest of parasites."[65]

Even with his considerable expertise, Ross required at least two hours to search for the parasite cells within all the tissues of a single mosquito.[65] To put this in perspective, Ross pointed out that at 1,000x magnification, "a mosquito appears as large as a horse."[65] And he had to compare many mosquitoes fed on malarial patients with those fed on healthy people. "I had no clue," he wrote, "as to the form and appearance of the object which I was seeking for; nor was I even sure that the kind of insect under examination was amenable to the infection at all. I was looking for a thing of which I did not know the appearance in a medium which I did not know contained it."[65]

Ross and Manson reasoned that once mosquitoes delivered the parasite to water, that water would be infectious to man. To test this, in 1895 Ross placed mosquitoes that had fed on malarial patients into jars with water until they died and theoretically infected the water.[65] Native Indian volunteers then drank the water to see if they would become infected with malaria.[87] Ross, unlike many of his contemporaries, considered the ethics of human experimentation. "The experiment was justifiable," he wrote, "owing to the slight degree of illness usually produced by malarial fever in natives when properly treated."[65] Three out of twenty-two subjects demonstrated a slight reaction to the malaria water, which Ross found inconclusive and possibly due to the fact that subjects were mostly from the lower castes of Indian society and may have used their participation fee for excessive drinking of liquor, which, he thought, could induce recurrence of a former malaria infection.

Ross came to the view that mosquitoes must transmit malaria from person to person, rather than from person to marsh water. He knew that mosquitoes bred in stagnant water, and that provided the connection to marshlands.[65] Ross tried various experiments to clarify the mosquito connection. He even had mosquitoes of all the suspect species feed on a patient who had three varieties of malaria in his blood, and then they were "applied in considerable numbers" to his volunteer, the assistant surgeon of a hospital in Bangalore.[65] But this fellow remained uninfected.

Meanwhile, Ross found himself distracted by a rivalry with Italian scientists (chiefly Amico Bignami, Giuseppe Bastianelli and Giovanni Battista Grassi) who attempted to shoot down his findings. Ross also felt harried by his official duties fighting a cholera epidemic, and discouraged by the continual negative outcomes of his malaria experiments. "Towards the end of my stay in Bangalore," he wrote, "as failure followed failure, I was naturally forced to reconsider the whole basis of my work."[65] But Manson put the screws on Ross to be quick about proving the mosquito hypothesis. The French (including Laveran) and the Italians were hot on his trail, "so for goodness sake," Manson wrote to Ross, "hurry up and save the laurels for old England . . . the fat is now in the fire and it is going to burn quickly."[88]

Ross decided to move his research operation to the area of India with the highest incidence of malaria. His request was denied by the Indian government, which felt that Ross's role as a medical officer was too important, given the outbreak of the Afridi War.[65] So

Ross used two months of leave that he had accumulated and his own funds to investigate malaria in the Nilgherry Hills (the Nilgiri Mountains of southern India). Ross took precautions: he slept at a rest house at 5,500 feet above sea level and only visited lower elevations during the daytime. But after just one excursion to the malarial area he became ill; he self-treated with quinine and recovered two weeks later. He found that nearly everyone in the area was infected, though mosquitoes were scarce. During his investigations he found a single individual of a new species of mosquito—he did not yet know that it was the one he was looking for.

Upon returning to duty at his military base, Ross found another species of mosquito similar to the new species he had found in the Nilgherry Hills.[65] Both, it turned out, belonged to the genus *Anopheles* (derived from the Greek, "good for nothing"). Ross, unfamiliar with mosquito taxonomy and unable to access any literature on the subject, did not know this; he worked strictly by describing the appearance and behavior of the different species without having associated names. He called them "dappled-winged mosquitoes."

These *Anopheles* mosquitoes were abundant and the incidence of malaria among the troops high, but he could find no larvae for testing.[65] He set out to repeat his experiments: larvae of the previously tested species were again collected and raised to adulthood, fed on malarial patients (all "trained to submit to mosquito bites—a matter often of some difficulty with the superstitious natives of India"), and then dissected at various time points, all with negative results.[65] Ross studied these mosquitoes with even more careful attention to every organ, to the feces, and to the contents of the intestines. Ross reported that the work was so grueling that he could scarcely see at the end of each day; his microscope decayed as his sweat rusted the screws that held it together, and the eyepiece cracked. "Swarms of flies persecuted me at their pleasure," he wrote, "as I sat with both hands engaged at the instrument."[65]

Finally, in August 1897, Ross made his critical discovery.[89] An assistant found a few specimens of *Anopheles* larvae; these were hatched and fed on malarial patients. After some botched dissections, two surviving mosquitoes were available for inspection. When Ross examined the first of these, he found nothing, but just as he was about to give up, he discovered pigmented cells in the wall of the insect's stomach.[65] The second surviving mosquito had the same cells in its stomach wall. "These two observations solved the malaria problem," he wrote. "They did not complete the story, certainly; but they furnished the clue. At a stroke they gave both of the two unknown quantities—the kind of mosquito implicated and the position and appearance of the parasites within it. The great difficulty was really overcome; and all the multitude of important results which have since been obtained were obtained solely by the easy task of following this clue—a work for children."[65]

Ross was elated. "The secret spring had been touched," he wrote, "the door flew open, the path led onward full in the light, and it was obvious that science and humanity had found a new dominion."[65] The next critical step was to study the life history of the parasite. With his experienced staff in place, sick patients accustomed to experimentation in the hospital, and a newly discovered breeding source for *Anopheles* mosquitoes, Ross was sure he could complete this task in a few weeks. But, without warning or explanation, the government transferred him to a remote outpost a thousand miles away. "It would be difficult for others to understand," he wrote, "the effect of this cruel blow."[65]

Meanwhile, Ross's Italian rivals pilloried him in scientific journals, finding ways to nit-pick his findings and thereby cast doubt upon them as they raced him to find malaria's mode of transmission. "I did not suspect that every line of mine," wrote Ross, "even in some of my private letters, would be subjected to a minute and unscrupulous analysis in the hope of finding discrepancies which would serve to discredit my observations. Every possible artifice has been used for this purpose by the very men who learnt all they knew from these very publications."[65]

Finally, after five months of frustration, Ross was able to return full time to his malaria research. Manson had used his influence with the government of India and the Indian Medical Service to get

Ross a year-long post for the study of malaria.[65] But now Ross was stymied again—this time not by bureaucracy, scientific rivalries, or a lack of the proper mosquito, but rather by riots. Bubonic plague was raging in India; just before Ross arrived at his new post, the Indian government attempted to inoculate the Calcutta population with an experimental plague preventive, or prophylactic. Serious riots ensued, "during which many of the Europeans had felt themselves obliged to go about armed with revolvers."[65] "The ignorant populace," wrote Ross, "thinking that the British were trying to inoculate them with and not against plague, flew into paroxysms of terror at the very sight of a European *hakim* (physician), while anything remotely resembling inoculation made them frantic."[65] Ross was barred from using malarial patients in hospitals for his research.

Eventually Ross was able to bribe some malarial Indian beggars to participate in his studies. "But when I proposed to prick their fingers in order to examine their blood," he wrote, "they generally left their money, took up their crutches, and fled without a word!"[65] So Ross, upon Manson's urging, turned to studying avian malaria, which is remarkably similar to human malaria. Manson pointed out that by infecting birds rather than human volunteers, Ross need not worry about "accusations of homicide."[88]

Ross let captive-raised mosquitoes feed on infected larks, sparrows, crows, and pigeons, and confirmed that mosquitoes transmit malaria.[65] He also found that if he fed the mosquitoes additional blood meals, he could keep them alive for a month rather than just a few days. This was critical: it showed that mosquitoes could easily live long enough to transmit disease from person to person; it allowed Ross to work out the detailed development of the malaria parasite; and it similarly facilitated the discovery by others of the manner of infection of yellow fever.

Ross hastily prepared a detailed report of his findings, which concluded with the statement that "these observations prove the mosquito theory of malaria as expounded by Dr. Patrick Manson."[65] Yet once again British bureaucracy obstructed Ross, and he was barred from publishing his findings without the permission of the secretary of state for India. So Ross asked Manson to publish the results for him. This publication finally drew positive attention to Ross's research, "to which," he wrote, "previously little credence had been attached."[65] Manson also felt the sting of criticism and wrote in one of his articles detailing Ross's findings, "I have been stigmatized as a sort of pathological Jules Verne, and hinted at as being governed by 'speculative considerations,' and as being 'guided by the divining rod of preconceived idea.'"[90]

Even with Ross's newfound recognition, and the critical importance of Ross's work (the mortality rate in India from malaria was estimated at ten thousand people per day), the government refused Ross's pleas for an assistant.[65] Although Laveran, Manson, and other experts accepted Ross's work as proof of the mosquito theory, the government would not support Ross until his work had been "confirmed." Ross estimated that the lack of assistance delayed his further work by more than a year, and the deployment of appropriate preventive measures in India by several years. "The truth is," he wrote, "that for some inexplicable reason men will never recognize the transcendent importance of investigation into the causes of those great diseases which destroy them."[65]

Nevertheless, Ross was able to press on with his research. In July 1898 he discovered that the spores of the malaria parasite collect in the salivary gland of the mosquito.[65,91] When the mosquito bites its victim, the spores pass with the saliva into the blood, causing malaria infection. He then demonstrated this experimentally by allowing malaria-containing mosquitoes to feed upon healthy birds and thereby pass on the parasite.[91] "The exact route of infection of this great disease," he wrote, "which annually slays its millions of human beings and keeps whole continents in darkness, was revealed."[65] He relayed his momentous finding to Manson by telegram (Ross and Manson exchanged more than two hundred letters between 1895 and 1899).[79,92]

At this point, Ross had just one more critical task to perform: to demonstrate that humans could be infected by malaria just as birds are. But once again, Ross was thwarted by his job, as he was

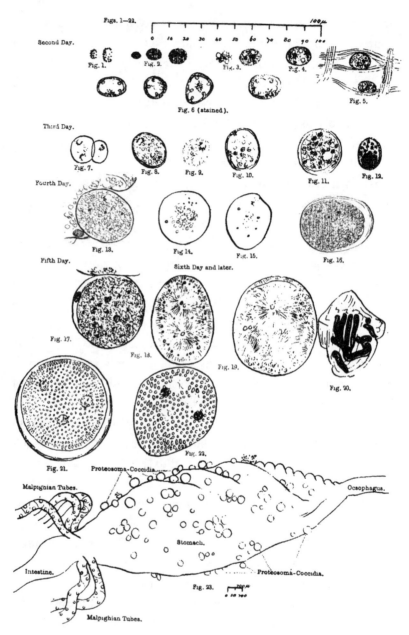

Fig. 2.1. Ross's sketches of developing malaria parasites in the mosquito, including the mosquito's stomach embedded with six-day-old parasites.[65]

assigned to work on a different disease. Many of Ross's critics then rushed to analyze his specimens, which he had shipped to England and France; they not only replicated Ross's findings but also claimed credit for them.[65] Ross's Italian rivals gained credit for Ross's work for some years, though Ross found their experiments "hasty and unreliable"—"misled by fondness for hypotheses," the Italians, he wrote, "fell into fundamental errors."[65] The credit they received provoked Ross. "Discovery is discovery," he wrote, "the determination of parallel facts, the filling in of details, the publication of pretty illustrations, and the furnishing of formal proofs of matters which are already certain, are useful—but do not constitute discovery."[65]

One of the Italians, Amico Bignami, successfully infected humans with infected mosquitoes in late 1898.[93] He was considered by many to have provided the first experimental proof of malaria transmission by mosquitoes to humans, though Ross considered this outcome a certainty, based on his earlier demonstration of mosquito infection of birds. "Bignami's experiment was merely a formality," Ross wrote, "of which the success could already be foretold with confidence."[65]

Late in 1898, Ross was permitted to return to Calcutta and his malaria research. But his energy was spent by relentless work and stress, and his health declined.[65] "The labour," he wrote, "the disappointments, even the successes, of the long and anxious investigations of a single subject had been too much for me."[65]

Although his own government took little interest in his work, others did, and soon Koch and other prominent scientists were actively engaged in follow-up studies.[65] Koch was the first to replicate Ross's findings of mosquito transmission, and Koch, like the Italians, also claimed credit for proving the mosquito hypothesis.[65,88] But Koch also supported Ross over the Italians in the contest for the Nobel Prize.[88]

In 1900, Manson fed *Anopheles* mosquitoes (infected in Italy with a benign form of malaria and shipped to England) on an unusual volunteer, his eldest son, Patrick Thurburn Manson.[58,94] Manson's son developed malaria, was treated successfully with quinine (just a few days before passing his final medical exams), and then had a recurrence of malaria nine months later, followed by a second recurrence during a shooting holiday—the first experimental documentation of recurrence.[79,95,96] Although this experiment replicated what the Italians had already done, it convinced many skeptics of the mosquito hypothesis because it was done in England, where there was no malaria, so accidental exposure outside of the experimental trial could not have occurred. Manson's son died from a gun accident on Christmas Island two years after the malaria experiment; he was only twenty-five.[58,96]

Manson also performed an exclusion experiment in a malarial part of Italy; here his volunteers stayed in a mosquito-proof hut at night and remained malaria-free while all their neighbors suffered from malaria.[94] The Italians performed a more extensive demonstration in 1900 in which 113 railroad workers were protected from mosquitoes (none contracted malaria), while 49 out of 50 unprotected workers fell ill.[97] Koch continued with his amazing career, and discovered, in the Valley of Ambarawa in Java, that epidemic malaria often struck down children but not adults, which meant that adults had acquired immunity.[98]

The subsequent findings of Koch and others that native children are the source of malaria parasites transmitted to European colonists in the tropics led to efforts by the British, Germans, French, and Belgians to segregate Europeans, especially in tropical Africa— even to the extent of destroying native huts in the vicinity of European homes.[99] "Segregation of Europeans at a distance from all natives," wrote the British team studying malaria in Nigeria, "offers itself now as the only measure by which absolute freedom from the disease can be guaranteed."[99] Koch also discovered in German East Africa that quinine killed the malaria parasite, and hence could be used to facilitate German colonization.[56] Other colonial powers also combined quinine treatments with segregation policies.

Ross was unable to secure additional work from the government to study malaria, so he reluctantly returned to England. Before leaving, he advised the Indian government, based on his research, to prevent malaria through the use of mosquito nets and through efforts to eliminate *Anopheles* mosquitoes.[65] His advice was ignored. But Ross understood the importance of his results—not just that mosquitoes transmit malaria from person to person, but that a particular type of mosquito that breeds in stagnant water is the perpetrator. Ross wrote excitedly of the implications for eradicating malaria by exterminating *Anopheles* mosquitoes: "What a weapon for good was now placed in our hands!"[65]

With newfound status and resources as the first lecturer at the Liverpool School of Tropical Medicine, Ross set out in August 1899 to complete his critical studies of malaria in the West African colony of Sierra Leone. "I had at last [like Odysseus in Homer's *Odyssey*] come to my Ithaca," he wrote, "after many mischances sent by many opposing deities."[65] Ross needed only a couple of weeks to find infected *Anopheles* mosquitoes, identify the malaria parasites in them, conduct feeding experiments, and establish that *Anopheles* mosquitoes only breed in stagnant pools of water.[100,101] "If the dangerous mosquitos prove to be confined to the genus *Anopheles*," Ross wrote, "the problem will be much simplified, and it will be advisable to declare war against the whole genus. The larvae of this genus can be distinguished by any intelligent European."[85]

The presence of *Anopheles* larvae in pools was sufficient to identify which pools must be dealt with.[65] Ross and his team determined efficient public health principles that should be employed in tropical regions: drainage of *Anopheles* breeding pools or their treatment with "culicicides" (pesticides that target mosquitoes), segregation of Europeans, protection of public buildings with mosquito screens, isolation of malarial patients, and individual use of mosquito nets.[100] Ross also attempted to change the name of the disease from "malaria," which connoted a false mode of transmission, to "hæmamoebiasis," referring to the amoeba-like pathogen and its infection of the blood.[101] The term did not catch on. He also suggested the term "gnat-fever," as he preferred the old English word for "mosquito."

Malaria spread with unprecedented speed as infected people and mosquitoes were rapidly conveyed by emerging technologies.

European progress in tropical colonies, such as those in Africa, could be measured by the expanding railway network and was accompanied by increasing malaria rates. Railway engineers, in particular, suffered a great deal of malaria.[101] When track was laid, regular pits were excavated along the line to afford material for the raised track. Mosquitoes were thus provided with ideal breeding habitat that cut across otherwise inhospitable terrain and a ready supply of men for their blood meals. Furthermore, the malaria parasite traveled in infected individuals on the growing network of rail lines throughout the world to infect new territories.[56]

Ross described the impact that malaria had on the fortune of nations. "Very malarious places cannot be prosperous: the wealthy shun them; those who remain are too sickly for hard work; and such localities often end by being deserted by all save a few miserable inhabitants."[78] He emphasized that malaria rendered "the whole of the tropics comparatively unsuitable for the full development of civilization."[78] Ross was awarded the Nobel Prize in 1902. In his acceptance speech, he pointed out that malaria "haunts more especially the fertile, well-watered and luxuriant tracts—precisely those which are of the greatest value to man. There it strikes down, not only the indigenous barbaric population, but, with still greater certainty, the pioneers of civilization, the planter, the trader, the missionary, the soldier. It is therefore the principal and gigantic ally of barbarism. No wild deserts, no savage races, no geographical difficulties have proved so inimical to civilization as this disease."[65]

Ross found the incidence of malaria in Africa to be particularly appalling. "We may almost say that it has withheld an entire continent from humanity," he said, "the immense and fertile tracts of Africa; what we call the dark continent should be called the malarious continent; and for centuries the successive waves of civilization, which have flooded and fertilized Asia, Europe, and America, have broken themselves in vain upon its deadly shores."[65]

Following his stunning scientific achievements, Ross vigorously promoted antimosquito campaigns and stressed the need to develop an improved and inexpensive insecticide to target mosquito larvae. "It must be some cheap solid substance," he wrote, "which kills larvae without injuring higher animals, and which, when sprinkled on depressions in the ground, will render the pools which form there uninhabitable for the larvae for a long time."[101] Yet here again, government inertia sapped his strength.

Ross recommended draining mosquito pools or treating them with the best available culicicide at the time, kerosene oil.[101] Others suggested using olive oil mixed with turpentine, iron sulfate, tar, lime, or peppermint oil. The British government informed Ross that mosquito eradication was impossible. Yet in towns in Egypt and Cuba where this was tried, researchers recorded an 80 percent decline in malaria. In Havana, the mosquito eradication efforts of 1901–2 also radically reduced the incidence of yellow fever, which had plagued the city continuously since 1760.[65] "The struggle over this matter," Ross wrote in 1902, "has been almost as severe as that over the original problem; but it is now drawing to a close."[65]

Despite successes in Egypt and Cuba, the available pesticides were inadequate. The British malaria expedition to Nigeria in 1901 reported that fumigation of homes with pesticides "is more likely to expel the European rather than the mosquitoes."[99] An important addition to the chemical arsenal came with the development of a paint pigment called "Paris green," or copper aceto-arsenite, into an insecticide used to kill mosquito larvae. Paris green had been used in 1868 as the first large-scale insecticide when it was employed against the Colorado potato beetle in the United States.[102] Paris green could be augmented with pyrethrum, a natural pesticide derived from chrysanthemum flowers, which killed adult mosquitoes.[56] But pyrethrum only killed for a short time and so was impractical on a large scale.

Unfortunately for the Manson-Ross collaboration, the two men's relationship soured after Manson criticized Ross's clinical skills as a physician (and after Ross downplayed the critical role Manson played in his discoveries); Ross ended up suing Manson for libel.[88] Despite all of his troubled relationships with colleagues and competitors, or perhaps because of them, Ross hoped for a brighter

Fig. 2.2. Photograph taken by the 1901 British malaria expedition to Nigeria, showing *Anopheles* larvae in different stages of their escape from the ova.[99]

future for medical research. "No form of enterprise is of such transcendent importance to humanity in general," he wrote, "as the investigation of disease—the principal enemy of every man . . . My labours will be abundantly repaid if earnest students in this field of science receive, in the future, in consequence of this narrative, a little more assistance than was given to me."[65]

[**3**]

Black Vomit
(1793–1953)

When it was evening I wished for morning; and when it was morning, the prospect of the labours of the day, caused me to wish for the return of evening.—**Dr. Benjamin Rush**, Philadelphia, 1794[103]

The story of "black vomit," or yellow fever, mirrors that of malaria. The disease arose in Africa, where it thwarted European efforts at colonization.[104] It traveled with soldiers, merchants, colonists, and slaves to the New World,[105] and then decimated the indigenous peoples of the Americas.[104] The greater immunity of African slaves increased their value, and hence yellow fever further entrenched the institution of slavery. Yellow fever also shaped the political landscape of New World colonies, perhaps best illustrated by its effects on the nascent United States.

The 1793 slave revolt in Saint Domingue (later Haiti) forced French colonists to flee. Some, infected with yellow fever, arrived in Philadelphia that summer, along with castaways hiding in the ships' water casks: the larvae of *Aedes aegypti*, the mosquito that transmits yellow fever. Fever broke out in Philadelphia that August, and by early November, 10 percent of Philadelphia's population had perished.[106]

Secretary of the Treasury Alexander Hamilton and his wife contracted the illness on September 5, and other prominent residents fell ill as well.[106] People with financial resources needed no more convincing, and 40 percent of the population bolted from the city. Those who remained were the poor, the sick, and those who bravely stayed behind to help others. As news of the calamity spread, other East Coast towns and cities pledged to prevent entry by anyone or anything from Philadelphia. One Philadelphia woman, upon trying to enter Milford, Delaware, was tarred and feathered, while residents of a nearby town sank an approaching ship that originated from Philadelphia.

Most of the city's doctors and other health care workers fled. The newly created Free African Society, a group of former slaves, was the only organization that volunteered to help the sick by serving as nurses, though after the crisis was over they were blamed for price gouging and stealing.[106] In his best-selling book printed as the epidemic waned in November 1793, Matthew Carey disparaged the black nurses, who themselves suffered significant mortality. "The great demand for nurses afforded an opportunity for imposition," he wrote, "which was eagerly seized by some of the vilest of the blacks. They extorted two, three, four, even five dollars a night for attendance, which would have been well paid by a single dollar. Some of them were even detected in plundering the houses of the sick."[107] Such racist accusations seemed out of character for Carey, who published antislavery articles, including some by black writers, and volunteered on the committee that oversaw city functions during the epidemic.[106]

The leaders of the Free African Society, Absalom Jones and Richard Allen, published a response to Carey in January 1794, entitled "A Narrative of the Proceedings of the Black People, During the Late Awful Calamity in Philadelphia, in the Year 1793: and a Refutation of Some Censures, Thrown upon Them in Some Late Publications."[108] This manuscript is the first one known in which African Americans challenged racist accusations.[106] In it, Jones and Allen, themselves former slaves, stated, "We feel ourselves sensibly aggrieved by the censorious epithets of many, who did not render the least assistance in the time of necessity, yet are liberal of their censure of us, for the prices paid for our services."[108]

High prices were indeed paid in some situations, but this was driven by sick people outbidding each other for the services of nurses; meanwhile, the Free African Society accrued financial liabilities from the assistance it provided. Jones and Allen asked what Carey would have demanded to nurse the sick and bury the dead had he not fled the city. They were dismayed at his branding of blacks as thieves, of his "unprovoked attempt . . . to make us blacker than we are," while neglecting to mention thievery by whites. "Many of the white people," they wrote, "that ought to be patterns for us to follow after, have acted in a manner that would make humanity shudder." For example, "a white man threatened to shoot us, if we passed by his house with a corpse: we buried him three days after." In many cases, neighbors and friends refused to help each other, black nurses who became ill attending to white patients were tossed onto the street, and orphaned children were abandoned. Noting that "an ill name is easier given than taken away," Jones and Allen provided a case-by-case review of heroic actions by black people.

This was not the first time that black people had been blamed for the consequences of yellow fever. In fact, some prominent physicians, noting the association between yellow fever and the slave trade, deduced that it was caused by "noxious miasmata" that come from "the negro body."[105] One physician wrote, "The organization of the negro, and the more extensive functions of the skin of this race as an excreting organ, give rise to the most offensive and foul state of the atmosphere when numbers of this race are confined in a limited space, and particularly in a humid and warm atmosphere. Indeed, nothing can be imagined more nauseous and depressing than the respiration of the air so contaminated."[105] He concluded that black people poisoned the air and thereby caused outbreaks of yellow fever among white people.

The most prominent Philadelphia physician who stayed to tend to the ill was Benjamin Rush, one of the signers of the Declaration of Independence, and a supporter of Jones and Allen in their effort to create the first independent black church in the United States.[109] Attending to the sick was a nightmarish task. Jones and Allen wrote of patients "raging and frightful to behold"; nurses attended to patients who were "vomiting blood, and screaming enough to chill them with horror . . . Some lost their reason and raged with all the fury madness could produce, and died in strong convulsions."[108] Rush wrote, "What medicine could act upon a patient who awoke in the night, and saw through the broken and faint light of a candle, no human creature, but a black nurse, perhaps asleep in a distant corner of the room; and who heard no noise, but that of a hearse conveying, perhaps a neighbor or a friend, to the grave?"[103]

Among those who fled at the start of the epidemic were the founding fathers of the United States residing in Philadelphia, which was the new nation's capital. George Washington wrote in a letter to James Madison that most of the chief players in the federal government suddenly had "matters of private concernment which required them to be absent."[110] Yet it only took a few days before Washington also decided it was time to abandon Philadelphia and return to his Mount Vernon estate. He left Secretary of War Henry Knox to run governmental functions, but Knox also took flight.

Washington's abandonment of Philadelphia created a constitutional crisis. Thomas Jefferson and James Madison (who would later marry Dolley Payne Todd, widowed due to the epidemic) argued that Washington could not call Congress into session outside of Philadelphia; this created a federal government shutdown at a time of chaos.[106,110] Their reasoning was sound. English kings had used the political ploy of calling Parliament into session in remote locations with no notice. So, as the fever abated, the founding fathers gathered in nearby Germantown, where Jefferson, Madison, and James Monroe all had to sleep on the floor or benches of an overfilled tavern.[106] When it was deemed safe, Congress convened in Philadelphia and passed a law allowing the president to assemble Congress outside of Philadelphia in cases of emergency.

Not only did the federal government collapse during the crisis, so also did the Pennsylvania state government and the Philadelphia city government, due to the death of some members and the flight of others. The mayor, Matthew Clarkson, stayed behind and created an illegal ad-hoc committee to run the city.[111] The committee was composed of men "scarcely known beyond the smoke of their own chimnies."[106] These men stepped into a governing role in the vacuum of leadership created by the epidemic.

The following year, in 1794, yellow fever revisited Philadelphia, and it did so again in 1796, 1797, and 1798.[106] The prospect of governmental functions perpetually subject to the effects of the disease deprived the city of its previous allure as the seat of government. Jefferson wrote in a letter to Rush that "yellow fever will discourage the growth of great cities in our nation," though in his opinion this was not necessarily a bad thing, since cities are "pestilential to the morals, the health and the liberties of man."[112]

John Adams also noted a positive side of the epidemic, which came just as tensions over the US relationship with France peaked and threatened to destabilize the young republic. The epidemic made the political conflict seem unimportant, and the fiery debate over relations with France fizzled away. "The coolest and the firmest minds," Adams wrote to Jefferson, "even among the Quakers in Philadelphia, have given their opinion to me that nothing but the yellow fever . . . could have saved the United States from a total revolution of government."[113]

Many influential people blamed the dirty Philadelphia water supply for the infection. Therefore, the city government decided that it needed a municipal water system. This system, constructed in 1799 by Benjamin Latrobe, became the nation's first municipal water supply.[114] Although contaminated water did not cause yellow fever, the construction of Philadelphia's water works led to the removal of numerous wells and cisterns, and therefore to the elimination of mosquito-breeding habitat.

Latrobe was the first professional architect in the United States. After completing the Philadelphia water system, he superintended construction of the Capitol building in Washington, DC, and subsequently served as architect of the replacement Capitol building after its destruction by British troops in 1814. Latrobe died from yellow fever in the 1820 New Orleans epidemic.[114]

Perhaps even more important than its effects on early US politics was the reversal of fortunes that yellow fever caused to France's American colonies. The French intended to stage troops in Haiti preliminary to the expansion of their North American empire. But Haitian slave revolts and the flight of French colonists set back those plans. Napoleon responded to the 1801 Haitian rebellion by dispatching his brother-in-law, General Le Clerc, with a formidable force.[106] The French troops slaughtered 150,000 Haitians as they wrested control from the former slaves, but then about 50,000 French soldiers, including Le Clerc, perished from yellow fever.[115] The few thousand surviving French troops fled, Haitian slaves regained their independence and created the world's first free black republic, and France lost its military stronghold in the West. Weakened, Napoleon fell back and sold the Louisiana Territory to the United States in 1803 for just $15 million.[115] The purchase more than doubled the size of the country; the new territory included what became fifteen states between the Mississippi River and the Rocky Mountains, extending from the Gulf of Mexico to Canada.

Later in the nineteenth century, the French effort between 1881 and 1889 to dig a canal across Panama collapsed due to the deaths of 30,000 workers from yellow fever and malaria.[106] The canal digging and the water containers used by the French created innumerable mosquito nurseries, and infected workers from around the world who had come to work on the canal provided ready blood meals and a source for disease outbreaks. This situation also opened an opportunity for the United States, which reinitiated the canal construction in 1904 and created an American canal territory that unlocked shipping between the Atlantic and Pacific Oceans. Why the United States succeeded where the French had failed less than twenty years earlier had everything to do with the discovery of the mosquito vectors of malaria and yellow fever.

But the United States also suffered a great deal from yellow fever, which not only brought Philadelphia to its knees, but repeated its performance throughout the country for the next century.[116] Yellow fever hit Manhattan every year from 1791 through 1821, and again in 1858.[106] Here, Irish immigrants were blamed, and sick Irishmen were shuttled from newly docked ships to the Marine Quarantine Hospital on Staten Island. Such was the fear of yellow fever that on September 1, 1858, an angry mob a thousand strong expelled the sick and burned the hospital to the ground.[117] They also torched the house of the resident physician and a nearby hospital, and when those fires were extinguished by police and firefighters, the mob returned to finish the job.

During the nineteenth century, yellow fever also swept through Boston, Baltimore, Mobile, Montgomery, Norfolk, Portsmouth, Savannah, Charleston, and Jacksonville.[106] It killed 8,000 people in New Orleans in 1853 and 7,000 in Memphis in 1873 and 1878.[118] New Orleans experienced thirty-nine yellow fever epidemics over the course of the century. Yellow fever outbreaks occurred in southern port cities of the United States in all but seven years between 1850 and 1900; in one of those years, 1861, the South was spared from yellow fever due to the blockade by Union forces in the Civil War.[119]

Practical measures were employed to impede the transmission of yellow fever, but these efforts did little since they were ineffective against the pathogen or its vector. For example, the prevalence of yellow fever in the South led to the fumigation of southern mail with formaldehyde.[106] Some southern communities tore up railroad tracks and burned bridges to keep infected people out.[120]

Casualties during the American Civil War, like those in most other wars, were caused primarily by disease. This war also apparently led to the use of yellow fever as a biological weapon.[121] The Kentucky-based physician Luke Pryor Blackburn, according to a Confederate double agent, tried to initiate yellow fever outbreaks in northern cities by shipping contaminated clothing from yellow fever patients in Bermuda.[122] He used this same technique to try to assassinate President Lincoln through a package of contaminated clothing.

After the war, Blackburn treated victims of several yellow fever outbreaks in the southern United States. He was called a hero in the

1873 epidemic in Memphis because he was the only physician "who offered to perform the Caesarian operation, in order that baptism might be administered to the unborn infant, when the mother was dead or in a desperate condition."[123] In this he contrasted sharply with other physicians, whose "shallow sympathy in favor of the mother often led them to deprive a human being of that dual life which God destined for it." Blackburn was known to use his cane to strike a rival physician on the head and shoulders in public, after accusing him of interfering with his medical practice; this sort of behavior was much admired by his contemporaries.[123] He also promised to care for a ten-year-old girl, the sole surviving member of a family of twelve that he doctored during the epidemic. After parading her around his hometown of Louisville, and appearing with her in the press, he sent her to a Catholic boarding school, but never paid them a cent. By then he was the governor of Kentucky.

The name "yellow fever" derives from the skin color of its victims, while the synonym "black vomit" is self-explanatory. A colonial doctor describing yellow fever symptoms in Jamaica in the late eighteenth century noted that in the progress of the disease, "the yellowness increased fast; so that the whole of the body was frequently yellow as an orange, or of as deep a colour as the skin of an American savage; anxiety was inexpressible; vomiting was irrestrainable, and the so much dreaded symptom of vomiting of a matter resembling the grounds of coffee, at last made its appearance . . . the colour of what was thrown up, was often black as soot."[67] A physician in Philadelphia wrote that "the patient has constant retchings to vomit, and discharges a dark coloured flakey substance, sometimes in small quantities; at others, by pints, quarts, and even gallons are discharged . . . his extremities become cold, his skin shrunk, his body assumes a cadaverous smell, and his breasts, neck, face, arms, and frequently every part of his body, is of a deep yellow colour . . . and ere long the functions of life cease, and nothing but an offensive mass of putrid flesh is left."[124] Despite this poignant description of symptoms, the author of this 1804 treatise felt that a patient's countenance defied all powers of description, such that it "would baffle the pencil of Raphael or Hogarth, and mock the efforts of a Shakespeare to delineate."

Physicians were befuddled by fevers that were difficult to differentiate from each other, and this contributed to ineffectual treatment. Even after malaria was clinically separated from other fevers due to the intervention of cinchona bark, most of the other febrile illnesses were still in a primitive state of treatment. "Medical writers have wandered from conjecture to conjecture, for more than two thousand years," wrote a late-eighteenth-century physician, "and we do not yet perceive any prospect of these conjectures being nearer to an end."[67]

Many treatments were employed to fight yellow fever, including bleeding; using hot mustard baths; wrapping the body in a blanket soaked with tea, salt, brandy, or rum; administering opium and wine; and shaving the head and soaking the body in warm water before dashing a bucket of cold saltwater onto the victim's head.[67,125] Doctors administered mercury and poisonous plants to induce vomiting and diarrhea in order to purge the illness from the patient.[124] During the 1878 epidemic in Memphis, the local newspaper advised readers to stay cool, avoid bad whiskey, keep to the usual activities, and "be cheerful and laugh as much as possible."[106] A prominent physician advised that "drunkenness, late hours, the excitement of gambling, debauchery of any kind favors its spread and increases the mortality."[126]

Bitter disputes arose between physicians over the efficacy of favored treatments. Benjamin Rush preferred aggressive bloodletting, a cure that Rush employed after reading a physician's letter[127] provided to him by Benjamin Franklin[106] about its utility in the earlier yellow fever epidemics in Virginia. During the Philadelphia epidemic, one prominent physician (a Princeton classmate of Rush's) criticized Rush's extreme bloodletting, saying that he would draw enough blood to fill a helmet "with as little ceremony as a mosquito would fill himself upon your leg."[109] Another physician considered it "pernicious treatment" that "sent many of our citizens to another world."[106] Rush's response to such criticism was defiant. "The envy and hatred of my brethren have lately risen to rage," he wrote to his wife. "They blush at their mistakes—they feel for their murders—and instead of asking forgiveness of the public for them, vent all their guilty shame and madness in execrations upon the man who has convicted them of both."[128]

The demand for Rush's bleeding and purging cure was so strong that patients had to be bled in the street, and Rush enlisted the Free African Society to administer bleedings for eight hundred people.[108] Other physicians also considered bloodletting the treatment of choice. One wrote that blood "may be drawn with a liberal hand, not confined to ounces but to pounds."[124]

In Philadelphia, many preventive measures were employed, such as putting a handkerchief soaked in vinegar over the nose, or burning gunpowder to cleanse the air.[106] Some people smoked tobacco to ward off the illness, while others chewed on garlic, fired muskets in their home, or covered their floors with a layer of dirt. Some physicians of the time understood the meagerness of the tools at their disposal. "The ordinary resources of our art are feeble," wrote one, "and if good can be done at all, it can only be done by means, which in the common opinion of practitioners, border on rashness."[67] Late in the development of the disease, "our labour is often the same, as if we attempted to resuscitate a corpse." Nevertheless, the public should not resort to "psychical depression," but rather "remember that in the most malignant epidemics the mortality is rarely over twenty-five per cent; there are, in other words, three chances for recovery to one for death."[126]

With the advent of the germ theory of disease in the 1870s, many experts on yellow fever reasoned that it, too, was caused by a living organism, though they were frustrated by the seeming impossibility of finding it. "So long as it is intangible, imponderable, irrecognizable to any of the senses," wrote a yellow fever investigator in 1876, "we have no positive knowledge of the essential nature of the poison."[129]

Since medical practice of the day had no way to stop the progress of the disease once a person became infected, emphasis was placed on disease prevention. A Louisiana physician, working in the midst of repeated and terrible yellow fever outbreaks, wrote in 1878 that the goal was "to attack these germs, wherever existing, by some agents destructive to low forms of life, without being injurious to their habitat."[116] The disinfectants employed for this purpose of attacking "an invisible foe" were sulfurous acid gas, sulfate of iron, lime, and diluted carbolic acid.[116,126,130] Unfortunately, these expensive efforts did not work, because the pests that needed to be destroyed were not "wingless animalcula" transferred by infected clothing, but rather viruses transferred by mosquitoes.

In response to the 1878 epidemic that sickened more than one hundred thousand Americans and killed one-fifth of them, the US Congress initiated a yellow fever commission to investigate causes and discover solutions.[131] By then yellow fever outbreaks had occurred in the territorial borders of the United States at least eighty-eight times, with the first known outbreak occurring in Boston in 1693. Nearly all cases were due to importation from the West Indies on ships; ships were particularly dangerous because "yellow fever infection clings to them with wonderful pertinacity in spite of all the methods of ventilation, disinfection, and purification that have up to this time been called into requisition." The commission wrote that "yellow fever should be dealt with as an enemy which imperils life and cripples commerce and industry . . . To no other great nation of the earth is yellow fever so calamitous as to the United States of America." The commission reported that "yellow fever often fails to swell into epidemic prevalence" under certain conditions, but it could not make a positive association between cause and effect. "The discovery of this unknown factor in the generation of yellow fever epidemics," the commission concluded, "would be a great boon to humanity."

Aedes aegypti (1880–1902)

By direction of the Secretary of War, a board of medical officers is appointed to meet at Camp Columbia, Quemados, Cuba, for the purpose of pursuing scientific investigations with reference to the infectious diseases prevalent on the Island of Cuba. Detail for the board:

Major Walter Reed, surgeon, U.S. Army;

Acting Assistant Surgeon James Carroll, U.S. Army;

Acting Assistant Surgeon Aristides Agramonte, U.S. Army;

Acting Assistant Surgeon Jesse W. Lazear, U.S. Army.

The board will act under general instructions to be communicated to Major Reed by the Surgeon General of the Army.—**Special Orders No. 122, Headquarters of the Army**, May 24, 1900[132]

It has been permitted to me and my assistants to lift the impenetrable veil that has surrounded the causation of this dreadful pest of humanity and to put it on a rational and scientific basis . . . The prayer that has been mine for twenty or more years, that I might be permitted in some way or sometime to do something to alleviate human suffering, has been answered!—**Walter Reed**, in a letter to his wife Emilie, December 31, 1900[133]

As in the case of malaria, some astute observers noted the association of yellow fever with mosquitoes long before their ideas were taken seriously. Josiah Nott, working in Alabama, hypothesized in 1848 that the yellow fever poison and the poisons of other "miasmatic" diseases were spread by insects.[134] Nott pointed out that it was impossible to explain the patterns of epidemics based on a poison in the atmosphere. For example, "yellow fevers, in 1842 and '43 travelled from house to house for more than a month, as would a tax collector, and was just about as much influenced by the weather; for neither the fever nor the tax collector like to travel in rain, though they pay no regard to the direction of winds." Similarly, in 1854 the French scientist Louis-Daniel Beauperthuy suggested that mosquitoes spread filth through their bite, thereby causing yellow fever.[135] A commission tasked to review Beauperthuy's mosquito hypothesis "all but declared him insane."[132]

Although nearly everyone who heard Nott's suggestion discounted it, a Cuban physician, Carlos Finlay, found it worthy of experimental study and concentrated his work on the species *Aedes aegypti* (at that time known as *Stegomyia fasciata*). Finlay noted that *Aedes aegypti* was common in areas plagued by yellow fever, it sought multiple blood meals, and it was active at temperatures associated with yellow fever outbreaks.[136] Finlay credited Benjamin Rush's account of the Philadelphia epidemic with first giving him the idea that mosquitoes were to blame, since Rush noted that "moschetoes (the usual attendants of a sickly autumn) were uncommonly numerous."[103]

In 1880, Finlay fed mosquitoes on yellow fever patients and then on five healthy individuals (including himself).[136] One of his subjects developed mild yellow fever. His publication on the mosquito as the vector of yellow fever was panned by critics, who considered the disease to be transmitted only by human-to-human contact. Two more decades passed before the mosquito hypothesis was taken seriously.

Ironically, the human-to-human method of transmission had already been disproved nearly a century earlier by Stubbins Ffirth, a doctoral student in medicine at the University of Pennsylvania. During the 1802 and 1803 yellow fever epidemics in Philadelphia, Ffirth collected black vomit and blood from yellow fever victims and applied these bodily fluids to cuts he made in his arms and legs (he performed similar experiments with cats and dogs).[124] He also tried inhaling fumes from black vomit that he cooked on an iron skillet, manufacturing the vomit into pills that he swallowed, eating fresh black vomit and blood, and putting the vomit in his right eye. When these efforts failed to induce illness, Ffirth injected himself with blood, saliva, sweat, bile, and urine from yellow fever victims.

Ffirth declared in his 1804 thesis, based on these experiments, that yellow fever is not contagious. This he considered a most important result because "as soon as persons are taken with fever, they are deserted by their dearest friends and relatives; the wife shuns the chamber of her husband, the husband of his wife, children of their parents, and parents of their children; they are consigned to the care of a mercenary, unfeeling black, and perhaps drunken nurse."[124] Ffirth also felt that proving that yellow fever was not contagious would render obsolete quarantine laws, which greatly damaged the economy. As research would show a century later, Ffirth was only partially correct; his hunch was sound that he could not get infected from the various bodily fluids—with the exception of infected blood. Somehow Ffirth escaped infection from the blood injection experiments. Ffirth also unintentionally suggested the means to eradicate yellow fever, by destroying stagnant ponds and draining meadows. He had in mind creating a clean environment, not exterminating mosquitoes.

Finlay did not give up on his mosquito hypothesis during the years when no one took him seriously, and, in fact, he extended the hypothesis into what he believed would be a public health tool. Finlay thought that mosquitoes infected with yellow fever could be used to vaccinate healthy people in order to impart immunity to the disease.[137] Similar ideas had been expressed regarding malaria by King and Koch.[75] Finlay tried this with sixty-seven people and claimed that his mosquito inoculations provided immunity.[138] Four of these people developed yellow fever, and one of them died. He repeated the experiment with forty-nine Jesuit and Carmelite fathers, and compared them with thirty-two control fathers who were not inoculated with mosquitoes. Five of the control subjects died of yellow fever. "The contaminated mosquitoes appear to lose," he wrote, "either partially or completely their contamination after they have stung healthy subjects; whereas the contamination appears to become intensified by successive stings of the same insect on yellow fever patients."[138] Since all of Finlay's subjects lived in an environment replete with yellow fever, and since he did not know about either the period of time when a patient can transfer yellow fever to a mosquito (the first three days of infection) or the incubation period of the pathogen in the mosquito (ten to sixteen days before the mosquito is infective), his results were inconclusive.[139]

In 1900, the surgeon general of the US Army, George Sternberg, created the US Army Yellow Fever Commission in Cuba. Cuba had recently fallen into American hands in the Spanish-American War. The motivation for the creation of the commission was an epidemic of yellow fever among American soldiers during the Cuban campaign. The success of the US military occupation of Cuba and other tropical territories depended upon success in the battle against yellow fever. Sternberg appointed Walter Reed, a US Army doctor, to head the commission.

A hard-working and brilliant man, Reed completed his medical degree at the University of Virginia at the age of seventeen after just a single year of study, then graduated with another medical degree a year later from Bellevue Hospital Medical College in New York City.[140] He worked as a physician in New York City and subsequently as an assistant surgeon in the US Army. He served in many frontier posts and even treated the famous Apache leader Geronimo. At one point Reed rescued an abandoned and badly burned Indian girl, only four or five years old. He brought her back to health and kept her as a nurse for his children. Reed's biographer in 1906 wrote of this charitable act: "When this child was nearly a woman the savage Apache blood asserted itself and she ran away, after giving abundant evidence . . . that fifteen years of kindness, gentleness, and refinement had not modified the cruel and deceitful characteristics of her race."[140] After eighteen years serving the army's medical needs, mostly on the American frontier, Reed joined the Army Medical School in Washington, DC, with the rank of major.

Under Reed's leadership, the commission included three other surgeons: Jesse Lazear (only thirty-four years old), James Carroll, and Aristides Agramonte. Lazear knew of the discovery two years earlier that mosquitoes serve as vectors for malaria. He had also read Finlay's 1881 paper hypothesizing that mosquitoes transmit yellow fever. And other insect vectors of disease had recently been discovered, including Patrick Manson's finding in 1878 that mosquitoes transmit elephantiasis,[72] Theobald Smith and Frederick Kilbourne's discovery in 1892 that ticks infect cattle with red water fever,[141] and David Bruce's work in 1894 showing that tsetse flies spread African sleeping sickness.[142] Lazear recognized the faults of Finlay's experiment on yellow fever; it was inconclusive and needed to be replicated in a controlled fashion. But both Sternberg and Reed found the mosquito idea far-fetched or, as Sternberg put it, "a useless investigation."[143] Sternberg and Reed prioritized research on the bacterium hypothesis proposed by the Italian bacteriologist Giuseppe Sanarelli.

Sanarelli had just identified a bacterium in yellow fever patients and named it *Bacillus icteroides*, or "yellow jaundice germ."[140,144] Sanarelli experimented with numerous animals, including mice, guinea pigs, rabbits, dogs, cats, monkeys, goats, sheep, donkeys, and horses.[144] Sanarelli also injected the microbe into five people (who did not give their permission), of whom three died, which made Sanarelli, for a brief time, a scientific celebrity for identifying the cause of yellow fever.[106] "These few but very successful experiments," he boasted, "have been sufficient to illuminate with a truly unforeseen light all the pathogenic mechanism until now so obscure and badly interpreted."[144] Others considered his human experimentation to be criminal. Regardless, Sanarelli's accolades stopped pouring in when Reed's commission studied the blood from yellow fever patients and found that the bacterium was actually a type of hog cholera that had contaminated Sanarelli's blood samples.[132,137,145–47]

Sanarelli was by no means alone in his use of unwitting subjects for medical research. Almost a century earlier, Edward Jenner infected his own baby boy with swinepox and smallpox before developing his pioneering vaccination against smallpox; the baby grew into a child of compromised health, and died at the age of twenty-one.[143] Sternberg and Reed, in an 1895 collaboration, tested a smallpox vaccine on children in orphanages.[148] But they also supported self-experimentation, most notably illustrated by Sternberg's experiment in which he swabbed his own urethra (and those of three terminal patients) with gonorrheal cultures; he was unable to infect himself or the others.

With Sanarelli's theory dashed, Reed permitted Lazear to spend time on the mosquito hypothesis. The commission members agreed that human experimentation was now necessary to advance their research, and they felt ethically bound to include themselves as research subjects if others were to be put at risk.[149] Of the four commission members, however, only Lazear was initially available for self-infection. Reed had duties back in Washington, Agramonte was thought to be immune since he was born in Cuba, and Carroll was occupied with other duties on the island.

Lazear and some volunteers let mosquitoes that had fed on yellow fever patients bite their arms.[132] They did not get sick. Carroll was skeptical of the mosquito hypothesis, and he eventually offered his arm to one of Lazear's mosquitoes. This one proved to be infective, and Carroll became the first person to be experimentally infected, though the commission could not conclusively establish that the mosquito bite was the source of infection.[137,145] The second was a hapless soldier named William Dean who had the misfortune to walk into Lazear's tent at just the time when Lazear happened to have mosquitoes ready for use.[132] "You still fooling with mosquitoes, doctor?" said Dean, to which Lazear replied, "Yes, will you take a bite?" "Sure," Dean replied, "I ain't scared of 'em." Dean's case convinced the commission that yellow fever was transmitted by mosquitoes.[137] Carroll nearly died, but both men eventually recovered. Carroll felt embittered and wrote to his wife that Reed's absence at the time of these self-administered mosquito trials by the commission members was too convenient.[149] The commission decided to halt the experiments.

Then Lazear, either intentionally or unintentionally, was infected by a mosquito and fell ill.[132,137] His wife Mabel, recuperating from giving birth to their second child in Massachusetts, received a cable nearly two weeks later that stated, "Dr. Lazear died at 8 this evening."[149] She had not known that he had contracted yellow fever. She wrote to Carroll to ask about the conditions of her husband's death: "[I am] anxious to know more about these circumstances as to how Dr. Lazear contracted yellow fever. In a note from General Wood yesterday he writes that Dr. Lazear allowed a mosquito to

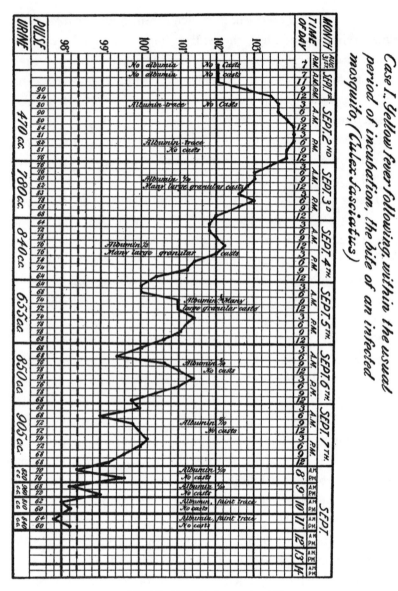

Fig. 3.1. Carroll's fever chart.[137]

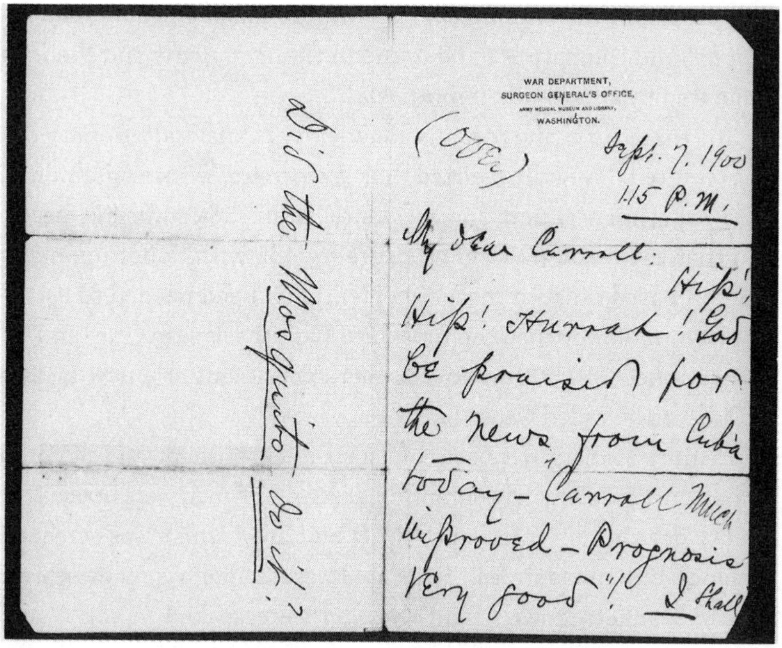

WAR DEPARTMENT,
SURGEON GENERAL'S OFFICE,
WASHINGTON.

(Over)

Sept. 7. 1900
1.15 P.M.

My dear Carroll:
Hip! Hip! Hurrah! God be praised for the news from Cuba today—"Carroll much improved—Prognosis very good"! I shall

Did the Mosquito do it?

Fig. 3.2. Reed heard of Carroll's recovery while in Washington, DC, and sent him this letter, in which he wrote, "My Dear Carroll: Hip! Hip! Hurrah! God be praised for the news from Cuba today—'Carroll much improved—Prognosis very good'!" On the back of the letter Reed wrote, "Did the mosquito do it?" From Hench Collection, University of Virginia.

bite him that had just bitten a yellow fever patient. Is it possible Gen. Wood could be mistaken—much as I know Dr. Lazear loved his work I can hardly think he could have allowed his enthusiasm to carry him so far."[150]

Reed deduced from Lazear's logbook that he had purposely infected himself, which would be considered suicide, and this in turn meant his family would be ineligible for a life insurance payout.[143] Lazear's logbook then vanished from Reed's office and only resurfaced fifty years later.

Reed hastily wrote and published an article detailing the commission's evidence that mosquitoes are vectors of yellow fever.[137] The article provoked criticism. "Of all the silly and nonsensical rigmarole of yellow fever that has yet found its way into print," read an editorial in the *Washington Post* on November 2, 1900, "the silliest beyond compare is to be found in the arguments and theories generated by a mosquito hypothesis."[106]

The three surviving commission members knew additional work was required. "We fully realized that three cases," wrote Agramonte, "two experimental and one accidental, were not sufficient proof, and that the medical world was sure to look with doubt upon any opinion based on such meager evidence."[132] Reed needed to establish beyond doubt that the mosquito hypothesis was true, and he requisitioned $10,000 to do a proper experiment at a new facility in Cuba, to be called Camp Lazear.

Healthy soldiers were selected for the experiment and put into quarantine.[132] Following quarantine, three of the men were housed for twenty days in the "infected clothing building," which was screened to block out mosquitoes. Their mattresses, pillows, pillowcases, sheets, blankets, towels, and clothing were soaked in the blood, black vomit, urine, sweat, and feces of yellow fever victims. On the first day, one of them retched when they opened a trunk full of soiled clothing. But they stayed the course and "passed a more or less sleepless night, in the midst of indescribable filth and overwhelming stench."[132] None of them, and none of their replacements in additional trials, contracted yellow fever, despite their constant handling of the soiled and vomit-soaked items.[151] As the commission expected, the disease did not spread through infected clothing, bedding, or air. This experiment demonstrated that tossing out or burning the belongings of yellow fever victims, as had been practiced for centuries, was a useless waste of resources.

Other volunteers were housed in the "mosquito building," with sterilized linens.[132] Agramonte transported mosquitoes that had fed on yellow fever victims in the hospital to the mosquito building so that they could feed on the volunteers. During one of his mosquito rounds, with the infected mosquitoes held in a cage tucked into his pocket, Agramonte's horse spooked at the sight of a steamroller and dashed down a hill out of control. His buggy flipped and sent him sprawling into the road. "The mosquitoes were quite safe,

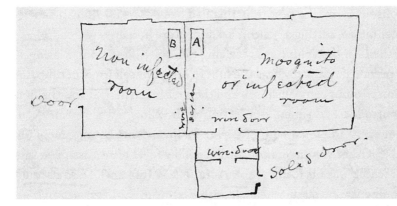

Fig. 3.3. Reed's diagram of Building 2 at Camp Lazear, sent in a letter to his wife on November 27, 1900. From Hench Collection, University of Virginia.

however," he wrote, "and upon my arrival at Camp Lazear I turned them over to Carroll for his subsequent care."[132] Two volunteers acted as a control group, experiencing the same conditions in the same building but with mosquitoes excluded from their room; they did not contract yellow fever.

The commission encountered difficulty finding volunteers for the infected mosquito trials. "Although several men had expressed a willingness to be inoculated," wrote Agramonte, "when the time came, they all preferred the 'infected clothing' experiment to the stings of our mosquitoes."[132] So Agramonte recruited Spanish immigrants, as they alighted on the docks, for light work at the camp. "Once brought in, they were bountifully fed, housed under tents, slept under mosquito-bars and their only work was to pick up loose stones from the grounds, during eight hours of the day, with plenty of rest between." During this time, Agramonte obtained their history and dismissed anyone who was a minor, was unhealthy, had lived in a tropical country, or had dependents. Then he offered one hundred dollars to the remaining men to be bitten in the mosquito house, with an additional hundred dollars if they contracted the disease. "Needless to say," wrote Agramonte, "no reference was made to any possible funeral expenses." The team employed an informed consent form, the first time this had been done for human experimentation, setting a benchmark for future human research.[149]

One of the volunteers receiving mosquito bites, John Kissinger, a private in the army, was the first to offer himself for experimentation without compensation, and the first to fall ill.[132] It was the first controlled experimental infection with yellow fever. He survived, but spent the remainder of his life an invalid and became mentally ill.[143] Other volunteers contracted yellow fever after Kissinger, one of whom was his former roommate, John Moran, who also rejected compensation simultaneously with Kissinger.[118] Moran was an Irish immigrant to the United States who had supported himself from the age of ten and joined the US Army Hospital Corps at the age of twenty-one.[152] When Moran's fever spiked, Reed said, "Mr. Moran, this is one of the happiest days of my life."[153] Moran, unlike Kissinger, made a full recovery.

Some of the volunteers from the infected clothing building also volunteered for the mosquito trials.[132] Although they did not get sick from contact with the soiled clothing and bedding, they did get sick when bitten by infected mosquitoes. "Rejoice with me, sweetheart," Reed wrote to his wife, "as, aside from the antitoxin of diphtheria and Koch's discovery of the tubercle bacillus, it will be regarded as the most important piece of work, scientifically, during the 19th century. I do not exaggerate, and I could shout for very joy that heaven has permitted me to establish this wonderful way of propagating yellow fever."[140]

When some of the Spanish immigrant volunteers were wheeled from the mosquito building to the yellow fever ward, other Spanish volunteers panicked and bolted.[132] Reed noted that "several of our good-natured Spanish friends who had jokingly compared our mosquitoes to 'the little flies that buzzed harmlessly about their tables,' suddenly appeared to lose all interest in the progress of science, and, forgetting for the moment even their own personal aggrandizement, incontinently severed their connection with Camp Lazear. Personally, while lamenting to some extent their departure, I could not but feel that in placing themselves beyond our control they were exercising the soundest judgment."[154] In Havana, people whispered of a limekiln full of the skeletons of Reed's volunteers, though none except Lazear had actually died, and the media reported that the Americans were injecting immigrants with poisons.[132]

The work of Reed's commission was not yet complete. With the success of the mosquito tests behind them, and knowing that the yellow fever pathogen must therefore be carried in the blood of its victims, the commission began blood injection experiments.[132] They withdrew 2 cc of blood from a yellow fever patient and injected it into a volunteer.[155] When he became sick, they withdrew 1.5 cc of his blood and injected it into another volunteer, who also fell ill. This experiment was performed on a few additional volunteers, and only one did not get sick.[132] The pathogen could pass from person to person in the blood, so humans served as a reservoir for epidemic outbreaks.

Another volunteer was injected with 0.5 cc of blood from a yellow fever victim, and he then fell ill. Four other volunteers who had had previous infections of yellow fever (Kissinger, Moran, and two others) were also injected with infected blood, and they did not get sick, which established that they had acquired immunity.[151] Reed decided that he would be the final volunteer, but in the last instant another man, John Andrus, stepped forward to take his place. Andrus was injected with infected blood, nearly died, but eventually recovered. Forty years later he lay paralyzed in the Walter Reed Hospital in Washington, DC, due, he was convinced, to the lingering effects of yellow fever on his spine.[143]

"Thus terminated our experiments with mosquitoes," wrote Agramonte, "which, though necessarily performed on human beings, fortunately did not cause a single death; on the other hand, they served to revolutionize all standard methods of sanitation with regard to yellow fever."[132] The utility of antimosquito measures was now obvious. "An epidemic can be stamped out in a community," wrote Agramonte, "by simply protecting the sick from the sting of the mosquitoes and by the extensive and wholesale destruction of these insects which, added to the suppression of their breeding places, if thoroughly carried out, are the only measures necessary to forever rid a country of this scourge."[132]

The clarity of the experiments drew scientific consensus behind Reed's commission. Even the *Washington Post* came to accept that mosquitoes serve as vectors of yellow fever, but the paper criticized the Yellow Fever Commission by asking, "Why not devote themselves to the eradication of the medium instead of killing more people by way of academic demonstration?"[143] A Philadelphia resident at the time of the 1793 epidemic had actually suggested killing mosquito larvae in water barrels,[106] but his idea was lost amid the noise of so many others.

Finlay was compared to Sir Patrick Manson, and Reed to Ronald Ross.[143] Finlay was nominated for the Nobel Prize seven times, though he was never awarded it; he was the most famous physician in Cuba. At a party in his honor, Finlay summarized his work. "Twenty year ago," he said, "guided by indications, which I deemed certain, I sallied forth into an arid and unknown field; I discovered therein a stone, rough in appearance, I picked it up and with the assistance of my efficient and faithful colaborer—Dr. Claudio Delgado, polished and examined it carefully, arriving at the conclusion that we had discovered a rough diamond. But nobody would believe us, till years later there arrived a commission, composed of intelligent men, experts in the required kind of work, who in a short time extracted from the rough shell the stone to whose brilliancy none can now be blind."[156]

Although Agramonte credited Finlay with developing the theory of mosquito transmission of yellow fever, he considered Finlay's experiments to be useless. Agramonte wrote, "The fundamental truth (the fact that mosquitoes transmit the infection), as conceived by Finlay, was so wrapped up in a mass of errors, hypotheses and speculations in his original contributions, that it remained hidden from the eyes of science until the subject received a thorough overhauling at the hands of the United States Army Board, when at last the grain was threshed out from the chaff; Finlay and Delgado's experimental cases formed part of the chaff."[157]

Shortly after the success of his commission, Reed grew concerned that he would not receive the credit he deserved. He wrote to his wife that he expected Sternberg to "write an article to say that for 20 years he has considered the mosquito as the most probable cause of yellow fever."[140] Sure enough, Sternberg claimed the credit for himself in an article published in 1901.[139] Sternberg also used the discovery to argue his case for a promotion in rank. "I beg leave to call attention to the fact," he wrote in his promotion application, "that the important discovery that yellow fever is transmitted by mosquitoes was due to my initiative. Without detracting in the least from the honor due Major Walter Reed and his assistants, who demonstrated this fact by a masterly series of experiments, the official records will show that this investigation was made upon my recommendation, and that the members of the Board were selected by me. I, also, gave personal instructions to the President of the Board, and pointed out to him the direction this experimental investigation should take."[143]

Research on yellow fever in Cuba continued. Finlay and his associate Juan Guitéras believed that a single mosquito bite produced only mild yellow fever, whereas multiple bites could produce the deadly form of the disease.[143] They attempted to immunize people with the bites of infected mosquitoes.[158] Reed was convinced by the commission's work that a single bite was enough. "After some poor devil dies," he wrote, "they may change their minds."[143] Guitéras proceeded to test forty-two volunteers using Reed's study methods, and three of them died; one was an American nurse who had volunteered herself for the research.[145,158] "I was very, very sorry to hear of Guitéras' bad luck," Reed wrote, "and can appreciate fully his mental distress over his loss of life—Perhaps, after all, the sacrifice of a few will lead to the more effectual protection of the many."[143] Guitéras's work demonstrated both that yellow fever strains differ in their deadliness and that he and Finlay were wrong about their single-mosquito-bite hypothesis. "The high rate of mortality in these cases," wrote Carroll, "removed all doubt as to the virulent nature of the attack induced by the bite of the mosquito, and the autopsy upon the first fatal case, which it was my privilege to perform, established the presence of the characteristic lesions."[145]

Carroll collaborated with Guitéras in his experiments in order to identify the pathogen that caused the disease. One impediment to their research was the now widespread knowledge that the pathogen was transmitted in the blood, either via mosquitoes or by injections. "The excitement following the fatalities that occurred later among Dr. Guitéras's cases," wrote Carroll, "rendered it exceedingly difficult for me to procure willing subjects."[145] Carroll was told by Guitéras that intentional infections with mosquitoes were no longer justified, but Carroll felt otherwise.

Carroll produced a couple more yellow fever cases using infected mosquitoes. He then filtered the patients' blood to remove bacteria, and injected the bacteria-free blood into volunteers who became ill.[145] Therefore, the pathogen, unlike bacteria, was small enough to pass through the filter, and far too small to see under the microscope. Carroll had demonstrated, for the first time, that a human disease was caused by a virus.[159] He could not state it as such since the word "virus" was not yet used in this context, nor was it possible to see these "ultramicroscopic" organisms. It was also the first demonstration of a virus transmitted by mosquitoes.

The Asibi Strain (1900–1953)

It seems to me that yellow fever will entirely disappear within this generation, and that the next generation will look on yellow fever as an extinct disease having only an historic interest. They will look on the yellow fever parasites as we do on the three-toed horse—as an animal that existed in the past, without any possibility of reappearing on the earth at any future time.—**William C. Gorgas**, 1911[160]

Reed's commission had discovered that a particular type of mosquito transmits yellow fever, but it did not discover the pathogen itself. The commission was able to disprove Sanarelli's hypothesis that *Bacillus icteroides* was the causative agent, but not without serious contention. At the 1901 meeting of the American Public Health Association, Reed presented the commission's groundbreaking work just two days after President William McKinley died from gunshot wounds to his chest and abdomen.[161] McKinley's anesthesiologist, Eugene Wasdin, arrived at the meeting from McKinley's deathbed, where he had incorrectly blamed the death on poisoned bullets.

Wasdin had served on the president's board appointed to study the Sanarelli bacterium, and he was convinced that it was responsible for yellow fever. "The fact that Dr. Reed states that the organism has not yet been discovered does not make that true," he said to the scientific assembly. "The organism has been discovered, and it is not inconsistent with Dr. Reed's demonstration of the transmission of the disease by the mosquito, to accept the organism of Sanarelli as the cause of yellow fever."[162] Wasdin considered mosquito transmission to be secondary to other routes of infection, such as from articles of clothing, and he stated that all of Reed's findings could be explained by the behavior of the Sanarelli bacterium. Reed, having just completed his trials, grew impatient. "It seems to me a waste of time," he responded, "with all due respect to Dr. Wasdin, who has labored so hard over this problem, to longer consider this bacillus as the cause of yellow fever."[162]

Not long after McKinley's assassination and the conflict with Reed, Wasdin displayed mental illness, and he died in an insane asylum in 1911 at the age of fifty-two.[163] Reed was considered for the Nobel Prize, but the first and second prizes, in 1901 and 1902, went to von Behring (for his work on diphtheria) and Ross (for his work on malaria); Reed died in 1902 from appendicitis at the age of fifty-one.[118] The general in charge of Cuban forces said at a memorial for Reed, "Hereafter it will never be possible for yellow fever to gain such headway that quarantine will exist from the mouth of the Potomac to the mouth of the Rio Grande."[118] Reed's grave in Arlington National Cemetery reads, "He gave to man control over that dreadful scourge Yellow Fever."

The human experiments in Cuba were halted (though both mosquito trials and blood injection experiments were replicated and refined by other researchers elsewhere),[145] and Carroll was denied a promotion advocated by Sternberg.[143] In 1907, a special act of Congress bestowed upon Carroll the rank of major, but his bout with yellow fever had so weakened his heart that he died later that year; he was fifty-three years old and left behind a wife and five children.[118,132]

The US government honored Reed, his team members, and the volunteers with a monument. This was fitting, given that the estimated number of yellow fever cases in the United States between 1793 and 1900 was half a million, and that in some years the economic loss from an epidemic exceeded $100 million.[118] With the problem solved by Reed's commission, Congress compensated the volunteers and the widows with a payment of $146 per month. Sternberg's replacement as surgeon general of the US Army criticized the compensation: "How discreditable appears this niggardly provision when compared with the action of the English Government which more than a century ago, when the purchasing power of money was far greater than at present, gave to Jenner, the discoverer of vaccination, grants amounting to 30,000 pounds sterling."[118] Ultimately, in 1929, Congress awarded the twenty-two volunteers (or their widows) a gold Medal of Honor in addition to an annuity of $1,500.[152]

Following Carroll's death, Agramonte was the only surviving member of the commission. Agramonte received no financial compensation, but continued as a professor of bacteriology and experimental pathology at the University of Havana. "Perhaps," he reflected, "it is thought that enough reward is to be found in the contemplation of so much good derived from one's own efforts and the feeling it may produce of innermost satisfaction and in forming the belief that one had not lived in vain. In a very great measure, I know, the thought is true."[132]

Cuba hosted both Finlay's experiments beginning in 1880 and the Yellow Fever Commission experiments in 1900. Cuba also benefited from the first fruits of that labor. Following the success of the Yellow Fever Commission, Reed's friend Major William C. Gorgas immediately led the eradication effort of *Aedes aegypti* in Havana.[164] His teams destroyed the breeding grounds and made widespread use of culicicides.

Aedes aegypti females deposited their eggs in the still water found in sewers, cisterns, barrels, gutters, ponds, outhouses, cans, and canals. Soldiers drained, smashed, or carted off anything small that held water, while larger water bodies were sprayed with kerosene, which prevented the mosquito larvae from breathing. Gorgas's teams found and treated 26,000 water deposits containing mosquito larvae in Havana by the time they issued their first report in March 1901. They installed screens on water tanks and stocked ponds with mosquito-eating fish. They quarantined yellow fever patients under mosquito netting and fumigated their rooms with pyrethrum, tobacco, and sulfur to kill infected mosquitoes. Gorgas's five-member teams could fumigate a large house in just two hours. After educating the public and making water containers mosquito-free, authorities began to impose ten-dollar fines on people whose property contained mosquito larvae, which motivated individual efforts as well.

Although the pesticides available at the turn of the century were limited to natural products, such as kerosene, pyrethrum, tobacco, and sulfur, Gorgas deployed them effectively. His mosquito eradication program followed a century and a half during which yellow fever invaded Havana on an annual basis.[147] In the previous forty-seven years, officials in Havana had recorded 35,952 yellow fever deaths. But within three months of the mosquito eradication program, Havana was essentially free of yellow fever and the malaria infection rate declined rapidly.[147,164] Ten months into the program, the number of water containers hosting mosquito larvae had dropped from 26,000 to 300. It was likely the most successful operation against disease vectors ever conducted.

When yellow fever struck New Orleans in 1905, the same aggressive antimosquito measures advocated by Gorgas were implemented by the US Public Health Service, with kerosene used as the pesticide. Such efforts were then quickly replicated elsewhere in the

United States and around the world, with successes in places such as the state of Texas, British Honduras, and Brazil.[132]

Also at the turn of the twentieth century, American interest in creating a canal across Panama peaked due to the Spanish-American War. At that time, Panama was part of Colombia. Reinforcing the Atlantic fleet with ships from the Pacific fleet required a journey around the tip of South America. To facilitate canal construction, the French (who wished to sell their rights and their railroad across Panama) and the United States conspired with a group of Panamanians (who did not want to lose the commercial benefits of a canal) in a revolution against Colombia. In November 1903, the United States recognized the new nation of Panama and immediately signed a treaty providing the United States with permanent and exclusive use of the Canal Zone. Panama received American money and military protection against a counterinsurgency from Colombian troops.

The following year, Gorgas applied aggressive draining and pesticide use in Panama to facilitate the construction of the canal.[165] While 30 percent of French workers had suffered the fevers induced by yellow fever and malaria during the failed French construction effort in the 1880s, only 2 percent of canal workers fell ill during the American construction.[106] It took only two years for Gorgas's team to eradicate yellow fever from Panama.

But the politics of Gorgas's program in Panama were dicey. Many people still ignored the work of Reed's commission and thought that Gorgas's antimosquito efforts were foolhardy. Consequently, Secretary of War William H. Taft and others lobbied President Theodore Roosevelt to sack Gorgas. But Roosevelt had seen firsthand what yellow fever was capable of during the Spanish-American War in Cuba, and so decided to seek the advice of a friend, Dr. Alexander Lambert, before acting upon Taft's recommendation.[166]

Roosevelt said, "They tell me Gorgas spends all his time oiling pools and trying to kill mosquitoes. Commissioner Shonts claims that he is not cleaning up Panama or Colon, that they smell as bad as ever, and recommends Colonel Gorgas's removal. The Secretary of War has gone over the matter and acquiesces in the recommendation."[166] The president's commission had reverted to the old ideas that yellow fever and malaria were spread by smells and filth. Lambert explained the science to Roosevelt and stressed that the canal could not be built without the antimosquito campaign. Roosevelt was impressed, and he kept Gorgas as the medical officer of the Canal Zone until the first ship navigated its locks in 1914.

By then, Gorgas had already been president of the American Medical Association for six years.[166] During World War I and the 1918 influenza pandemic he served as surgeon general of the US Army. In 1920, on his way to Africa to study yellow fever as part of a new Yellow Fever Commission, Gorgas stopped in London to receive the honors of King George V. He suffered a stroke while there, but received a knighthood from the king in the hospital, where he lingered for a few weeks and then died. His wife wrote, "It must have been a keen disappointment to be so suddenly halted in the midst of his labours, at the very moment the promise of fulfillment was near—the fulfillment of his dream to rid the world of yellow fever and make the tropics safe for the white man."[166]

Another member of Gorgas's 1920 Yellow Fever Commission, Adrian Stokes, joined the 1927 team sent by the Rockefeller Foundation to Nigeria "to attempt the isolation of the organism which causes the disease."[167] By then, investigations of infectious diseases had moved beyond human studies to effectively employ animal models. The 1927 team needed to develop an appropriate animal model for yellow fever, and in their search for the right species they injected blood from yellow fever patients into guinea pigs, rabbits, rats, mice, Gambian pouched rats, dogs, cats, goats, chimpanzees, and various species of monkeys.[168] Of these, the rhesus monkey proved to be susceptible to yellow fever and therefore a good animal model for the disease.

On June 30, 1927, the team took blood from a feverish man named Asibi (who survived) and returned to their lab in Accra (the Gold Coast, or Ghana), where they injected it into a rhesus monkey.[168] The monkey died, and Stokes's team performed an autopsy to confirm a diagnosis of yellow fever. Its blood was injected into another monkey, which also died, and this experiment was replicated in thirty monkeys, only one of whom survived. The team then passed yellow fever between monkeys through the bites of mosquitoes. Monkeys, it turned out, acted as a reservoir of the yellow fever virus, which explained how human epidemics could arise in urban areas adjacent to rainforest. Additionally, the team found that individual mosquitoes remained infective as long as they lived, and they could live longer than three months. Reed's team had earlier made the same important discovery, and noted that "we have for the first time an explanation of the fact, several times noted in the literature, that the contagion of yellow fever may cling for several months to a building that has been vacated by its occupants, or to the infected area of a town, even though this latter has been entirely depopulated."[155]

Stokes fell ill on September 15, 1927, with the Asibi strain of yellow fever, possibly acquired through accidental exposure of infected blood through a monkey bite wound on his hand.[169,170] From his hospital bed in Lagos, Stokes asked his collaborators to feed mosquitoes on him and then transfer the mosquitoes to their experimental monkeys, as well as to inject his blood into the monkeys; both experiments led to fatal cases of yellow fever in the monkeys.[171] He also insisted that his colleagues autopsy him should he die.[143] The infection proved fatal, and the autopsy revealed a positive case of yellow fever in what appeared to be the first documented instance of transmission through the skin.[172] Stokes's colleagues published their discoveries shortly after his death, and honored Stokes with first authorship.[168]

Four other Rockefeller Foundation researchers subsequently died of yellow fever during their investigations, including the famous Japanese bacteriologist Hideyo Noguchi.[169,173] Noguchi had spent nearly a decade leading researchers astray with his incorrect conclusion that yellow fever was caused by a bacterium which he called *Leptospira icteroides*.[172] Stokes had demonstrated that it was not the responsible pathogen; in his effort to provide a final proof of this, Stokes had a prominent physician and Noguchi advocate examine him on his deathbed.[171] Stokes asked the physician, "Are you ready now to agree, to admit that yellow fever is caused by a virus and not by *Leptospira*?"[171] The physician responded, "I believe you fellows are right. I don't have the explanation, but I think you have yellow fever and got infected in the laboratory with what you call a virus."

But Noguchi was persistent in his claims. Noguchi inoculated himself against *Leptospira* and thereby considered himself protected from yellow fever. Then he initiated his own studies in the Accra lab, only to find that he was indeed wrong. Carroll had been correct that yellow fever was not caused by a bacterium. Shortly after Stokes's death, Noguchi died in Accra of yellow fever. Researchers advanced two theories for how Noguchi became infected. One posited deliberate infection following the discovery of his monumental error, a scientific form of *seppuku*, or Japanese ritual suicide.[172] Another theory, considered more likely, held that infected mosquitoes escaped from experimental cages that were sloppily maintained in Noguchi's lab.

Not long after, researchers discovered that *Aedes aegypti* was not the only mosquito that could transmit yellow fever.[174] When slavers sailed to the New World, they transported both the pathogen (in infected slaves) and the vector (*Aedes aegypti*).[104] Slavers likely first introduced the yellow fever pathogen to the Americas in the mid-seventeenth century. Infected mosquitoes then passed the pathogen to people and monkeys in the Americas, and a permanent reservoir for the disease was established. Some of the mosquitoes native to the American tropics then also started carrying the disease. Worse still, *Aedes aegypti* is not only guilty of transmitting yellow fever - it infects people with two other major diseases as well: dengue fever and chikungunya (which means "to walk bent over" in Kiswahili), each caused by a virus similar to the yellow fever virus.[175,176]

The path to a vaccine also claimed its string of casualties. A key development came in 1930 when Max Theiler demonstrated that white mice could be used as an effective model organism for yellow fever investigations.[177-79] Compared to monkeys, mice allowed for more efficient research at a fraction of the cost. The next year,

Theiler protected mice from infection by inoculating them with blood serum from infected monkeys or humans. This gave scientists a powerful tool to map the occurrence of yellow fever in people and of jungle fever in monkeys (which leads to new yellow fever outbreaks in people). By transmitting the yellow fever virus from mouse to mouse, Theiler found that it became less dangerous and could be used to vaccinate monkeys.[179] The mouse vaccine was then developed for use in humans, and in 1938 Theiler's team developed a second human vaccine from the Asibi strain, taking advantage of a chance mutation that occurred in the lab (the 17D vaccine).[180,181] Theiler contracted yellow fever, though his was a mild case.[182] The US military inoculated its soldiers during World War II; although eighty-four people died from contaminated vaccinations, not a single American soldier contracted yellow fever.[183] A French team also developed an effective vaccine that could be delivered efficiently to large populations.[184] By 1953, 56 million Africans had received the French vaccination.[185] Eventually, the 17D vaccine was universally viewed as superior, and the French vaccine went out of production.[182]

In 1948, Theiler was nominated for the Nobel Prize by Albert Sabin, who later developed the live vaccine that stamped out polio from most of the world.[186] However, that year the prize went to Paul Müller "for his discovery of the high efficiency of DDT as a contact poison against several arthropods." Three years later, hours before the Nobel Committee's deadline for receipt of nominations, its chairman renominated Theiler and performed the evaluation of his own nomination. This nomination succeeded, and Theiler received the only Nobel Prize ever awarded for work on yellow fever, or for work on a virus vaccine. Also in the 1950s, researchers tracked down Asibi in the Gold Coast of West Africa, and the British Colonial Office awarded him a pension for his contribution to the development of the yellow fever vaccine.[172]

The science of yellow fever followed a convoluted path. First, visionaries such as Nott and Beauperthuy suggested in the mid-nineteenth century that insects, possibly even mosquitoes, transmit the disease. Finlay took this further by correctly stating in 1881 that a specific mosquito, now called *Aedes aegypti*, transmits yellow fever, and he initiated poorly controlled experiments to demonstrate this. Finlay was stung by two decades of ridicule before Reed's commission performed the critical and conclusive experiments. This proof that *Aedes aegypti* transmits yellow fever led Gorgas to immediately eradicate the disease from Havana, and similar massive mosquito elimination programs were initiated in Panama and other yellow fever hotspots. Meanwhile, Carroll established that the pathogen was smaller than bacteria, but its identity was still a mystery that lay beyond the reach of turn-of-the-twentieth-century technology. Stokes and his collaborators in 1927 established the link to monkeys and jungle fever outbreaks, and this allowed the isolation of the yellow fever virus in monkeys in 1928[187] (the genome of the virus was sequenced nearly sixty years later).[188] In 1930, Theiler and his colleagues created the mouse model of yellow fever, and eight years later they developed the 17D vaccine. The disease could then be attacked from all sides. Vaccination prevented human infection, and widespread use of pesticides and other anti-mosquito measures reduced the populations of the vector. Yellow fever was stamped out—for a time.

[**4**]

Jail Fever

(1489–1958)

Soldiers have rarely won wars. They more often mop up after the barrage of epidemics. And typhus, with its brothers and sisters,—plague, cholera, typhoid, dysentery,—has decided more campaigns than Cæsar, Hannibal, Napoleon, and all the inspector generals of history. The epidemics get the blame for defeat, the generals the credit for victory. It ought to be the other way round.—**Hans Zinsser**, typhus researcher, 1934[40]

Lice transmitted two epidemic diseases during the Irish Potato Famine—typhus and relapsing fever, both referred to by the Irish as "famine fever."[27] The Irish had a long, ugly history with typhus because it broke out whenever the potato crop failed. Remarkably, less than thirty years before the great famine, typhus infected 700,000 of the 6 million Irishmen.[189] Then it fell dormant until the sudden arrival of the potato blight.

During the famine, the louse's role in infection was unknown. It was not until the 1870s that Pasteur and Koch developed the germ theory of disease. During the following three decades, in perhaps the most remarkable period in the history of public health research, the responsible germs and the animal vectors were discovered for numerous infectious diseases, including malaria, yellow fever, and bubonic plague. But in 1900 the germ and vector of typhus still remained a deadly mystery.

Or rather, it was a mystery to all save Albert Freeman Africanus King. Not only was King prescient—though ignored—when it came to the mosquito hypothesis and malaria, so too did he predict, without notice, the mode of transmission of typhus and yellow fever. "With our present knowledge of the 'germ theory,'" he wrote in 1883, "one would hardly dare, even once, to plunge an inoculating needle into the blood of a yellow-fever or typhus-fever patient, whether living and comatose or recently dead, and then withdraw it and plunge it into his own blood or the blood of other persons, yet this is exactly what the mosquito is doing in nearly every yellow-fever epidemic, and what, perhaps, the flea is doing in the filthy jails and ships infested with typhus."[54]

The word "typhus" derives from the Greek *typhos*, which means "smoky or hazy" and describes the mental condition of infected individuals.[189] An older name for typhus is "jail fever" (or "jayl fever" or "gaol fever") due to its prevalence in crowded prisons. Although British law in the centuries before the famine sometimes listed more than two hundred capital offenses, including such crimes as shoplifting, horse theft, letter stealing, and witchcraft, more prisoners perished from typhus than from the noose.[190] Furthermore, prisoners often infected court personnel during trial, a phenomenon referred to as the "Black Assizes."

One notable example of a Black Assize was initiated by a prisoner named Rowland Jencks at Oxford in 1577.[40] Jencks, a Catholic bookbinder, was arrested for criticizing the government, profaning God, and avoiding church. "Considering the times," wrote bacteriologist Hans Zinsser, "he appears to have been a fellow of spirit and conviction."[40] The case aroused considerable public interest. Unfortunately, Jencks caused a typhus outbreak among the crowd who attended the trial. Subsequently, the judge (who had been the Speaker of the House of Commons), the sheriff, the undersheriff, all but two members of the grand jury, one hundred members of the faculty of Oxford College, and a few hundred other people died. Sir Francis

Bacon investigated the outbreak and determined it was caused by corrupted air.[191] Others blamed "papistical evil magic" released in town.[40] But the facts of the case, viewed through the lens of modern science, force a conclusion that "no inconsiderable number of the faculty of Oxford College were, at this time, lousy."[40] Jencks had his ears cut off, but survived for another thirty-three years working as a baker in the English College of Seculars.

Similar Black Assizes occurred at Canterbury in 1522, Exeter in 1589, Taunton in 1730, Lanceton in 1742, and at the Old Bailey in London in 1750.[40,192] In response, the celebrated eighteenth-century advocate for prison reform, John Howard, pushed through improvements in prison conditions that reduced the incidence of typhus.[40,190] In 1790, Howard himself died of typhus while inspecting a prison in the Ukraine.

Because typhus flourished during times of famine and in foul living conditions, epidemics were particularly common during wars and often determined the outcome. When the Spanish forces of Ferdinand and Isabella laid siege to Moorish Granada in 1489–90, they lost three thousand men to combat and seventeen thousand to typhus.[40,193] At the cusp of victory, the majority of the French force that attacked Naples in 1528 fell victim to a typhus epidemic; Pope Clement VII was forced to submit to Charles V of Spain, who thereby gained the crown of the Holy Roman Empire.[40] The hopes of Maximilian II, the Holy Roman Emperor, were dashed in 1566 when typhus tore through his eighty thousand troops amassed to attack the forces of Sultan Suleiman the Magnificent in Hungary. In the New World, that same year and the one following, typhus killed more than 2 million indigenous Mexicans. This was just one of the waves of epidemics brought by Europeans and enslaved Africans that decimated the indigenous peoples of the New World.[189]

Typhus ravaged the forces of the Swedish king Gustavus Adolphus and the military commander Albrecht von Wallenstein in the service of the Holy Roman Empire as they met for battle at Nuremberg in 1632; both armies retreated before fighting could break out, in yet another instance of epidemics driving the outcome of the

Thirty Years' War.[40] Besides typhus and bubonic plague, the Thirty Years' War gathered together the forces of dysentery, typhoid fever, diphtheria, smallpox, and scarlet fever;[40] in one notable example, the population of the German region Württemberg collapsed from four hundred thousand to just forty-eight thousand.[192]

The city of Prague fell to the French in 1741 because the defending Austrians lost thirty thousand soldiers to typhus.[40] Typhus defeated Napoleon's mighty army at the height of his power in 1812, altering the course of European history.[193] Out of half a million soldiers in Napoleon's invasion force, only eighty thousand reached Moscow and fewer than ten thousand returned to France.[192] Typhus and other epidemics were equal-opportunity destroyers during the Crimean War; between 1854 and 1856, typhus first plowed through the ranks of the Russians, then the French and the British, and then spread through the navies and merchant ships to land-based hospitals and into a general epidemic.[40] While approximately sixty thousand French soldiers were wounded or died in battle, about a quarter of a million were sickened or died of disease; similar results were felt by the British and Russians. "Typhus," wrote Zinsser, "had come to be the inevitable and expected companion of war and revolution; no encampment, no campaigning army, and no besieged city escaped it."[40] For those who contracted typhus and other infectious diseases during these centuries, their odds of survival were greatest if they did not procure the service of a physician, since medical treatments such as bleeding and nonsterile surgeries, as well as pathogens spreading through health facilities, typically worsened the patient's condition.

By the late eighteenth century, the pioneering Scottish physician James Lind concluded that typhus was carried by clothing and other materials.[40] He noticed that workers assigned to refurbish hospital tents died of typhus, and that it spread through the bedding on board ships. He pushed for fumigation to combat typhus, but the fumigants at his disposal, derived from the likes of tobacco, charcoal, vinegar, pitch or tar, and gunpowder, were ineffective. Lind made serious inroads against the disease, however, through sanitation efforts. He introduced the idea that medical staff should routinely change clothing. Although Lind was unaware of insect vectors of disease, his observant nature and intuition led him to the threshold of stopping typhus. Ultimately, the lack of an effective fumigant stymied his progress.

Clearly, typhus was one of the most important diseases to conquer. Rapid discoveries in the late nineteenth century of the causes of malaria, yellow fever, and bubonic plague created eager anticipation for the discovery of the cause of typhus. Identification of the responsible pathogen might present medical solutions, such as a vaccine. Identification of an animal vector might present a means of breaking the infection cycle through the use of pesticides. The vector was discovered before the pathogen by a man named Charles-Jules-Henri Nicolle.

Nicolle loved literature and the arts, but he heeded the directive of his father, a physician, to follow in his footsteps.[194] Nicolle received his medical degree from the Institut Pasteur in Paris in 1893, and then returned to his hometown of Rouen for a faculty position. There he found his progress thwarted by insecure employment, colleagues who did not accept his scientific ideas, and a loss of hearing that precluded the use of a stethoscope. So when his older brother turned down the position of director of the newly created branch of the Institut Pasteur in Tunis, Nicolle applied for the job in the hope of transferring to a more promising career.

Nicolle moved to Tunis in 1902; he was only thirty-six years old, and would hold the institute directorship until his death in 1936.[194] Nicolle determined that of all the infectious diseases ravaging North Africa, typhus was "the most urgent and the most unexplored."[195] He noticed that typhus swept through the community in seasonal fluxes during cool parts of the year, and that those most affected were the poor. Densely occupied prisons, asylums, and makeshift settlements acted like magnets for the disease, and it frequently killed hospital staff and doctors. The majority of Tunisian doctors contracted typhus, and one-third of them died from it.

The first time Nicolle scheduled research on a prison outbreak of typhus, in 1903, his stellar career was nearly cut short; just before joining his two colleagues, Nicolle coughed up blood and so canceled his participation. His colleagues spent the night in the prison, contracted typhus, and died.[195] "The fact that I was fortunate enough to escape contagion," wrote Nicolle, "in spite of frequent, sometimes daily contacts with the disease, was because I soon guessed how it spread."[195]

Nicolle studied the people at the entrance and in the waiting room of a local hospital for the poor.[195] He discovered that typhus patients infected others during the hospital admission process up to the point when they were washed and dressed in clean clothes. Meanwhile, medical staff who took away their dirty clothing often contracted the disease. "I asked myself," wrote Nicolle, "what happened between the entrance to the hospital and the wards. This is what happened: the typhus patient was stripped of his clothes and linen, shaved and washed. The contagious agent was therefore something attached to his skin and clothing, something which soap and water could remove. It could only be the louse."[195]

The lice hypothesis finally explained why war and famine almost inevitably led to typhus, and it made sense of the common names given to the disease over the centuries: "jail fever," "famine fever," "Irish fever," "camp fever," "ship fever," "hospital fever." When undernourished people were crowded together in filthy conditions, lice accelerated the pathogen's transmission into epidemic proportions.

To test his lice hypothesis, Nicolle requisitioned a few chimpanzees from Emile Roux, then the director of the Institut Pasteur in Paris. The day the chimps arrived, Nicolle inoculated one with blood drawn from an infected patient.[195] A day later, the chimpanzee lay drained, heated with fever and blemished with skin lesions.[194] Nicolle gathered lice from the sick chimpanzee and put them on another chimp, which also contracted typhus.[196] Chimpanzees were expensive, so Nicolle extended the experiment by injecting blood from the sick chimpanzee into a toque macaque.[195] After thirteen

Fig. 4.1. Charles-Jules-Henri Nicolle.

days, the monkey also came down with fever.[194] Nicolle then placed twenty-nine lice onto the fur of the sick macaque and gave them time to feed. He transferred the lice to other macaques, which also became ill, but after they recovered, they were immune against further infection.[194,195]

In September 1909 Nicolle announced to the French Académie des Sciences that typhus is vectored by lice, thereby opening the door to typhus eradication through delousing efforts.[197] He also succeeded in infecting guinea pigs, which, unlike primates, were cheap; this allowed for continuous research on typhus, which had previously been limited to periods of epidemic.[195]

Nicolle's evidence of lice transmission was confirmed a year later in Mexico by the American scientists Howard Taylor Ricketts and Russell Morse Wilder.[198] Also in 1910, Nicolle demonstrated that the pathogen reproduces in the louse's digestive tract, and that the feces of the louse infect the host.[195] Using these results as a guide, Henrique da Rocha-Lima discovered the responsible pathogen in 1916.[199] Rocha-Lima honored Ricketts and the Austrian scientist

Stanislaus Joseph Matthias von Prowazek,[200] both of whom had died of typhus while studying it, by naming the pathogen *Rickettsia prowazeki.*[40]

The human body louse lives in clothing and deposits its eggs in undergarments.[189] The eggs incubate for eight days before hatching, and then the nymphs undergo three molts during a two-week period to mature into adults. Adults leave the security of the clothing only to feed on their host's blood, which is their only nourishment. The louse contracts typhus from an infected person's blood, and the *Rickettsia* pathogen appears in the louse's excrement a few days later. There, *Rickettsia* may remain viable for months.

Lice prefer normal human body temperature, so they tend to leave febrile infected people to parasitize uninfected people, and similarly they abandon a corpse to seek a more suitable home. On the new host, the louse bites a small incision to draw blood, and it simultaneously defecates. Nicolle demonstrated that when a person scratches at the bite, the feces are rubbed into the wound, initiating a new infection.[195] Alternatively, infection may occur if the louse itself is crushed into the wound or the louse excrement is rubbed into the eyes.[189,195]

Nicolle's discovery of louse transmission had immediate practical importance. The public health department in Tunis began aggressive delousing efforts and, within a few years, succeeded in eradicating typhus from urban areas, mines, and even prisons.[195] Louse eradication efforts like those in Tunis were adopted all over the world, and so Nicolle's discovery saved countless lives.

Nicolle believed that he could create a vaccine by mixing the typhus pathogen with blood serum from typhus survivors.[196] He administered this concoction to himself and remained healthy. He next tried it out on children, due to their greater resistance to the disease. "You can imagine how frightened I was," Nicolle said, "when they developed typhus; fortunately, they recovered."[196] Nicolle's typhus vaccine was inadequate and so was not adopted. But capitalizing on his knowledge gained from experimenting with a typhus vaccine, Nicolle did produce an effective measles vaccine, using blood serum taken from children who survived the disease.[195]

In addition to his critical discovery that typhus is transmitted by lice, and his significant work on other infectious diseases, Nicolle described in 1911 how an individual could acquire and convey an infectious disease without displaying symptoms.[195] He discovered this process of "inapparent infection" in typhus-infected guinea pigs. Some guinea pigs appeared healthy even though they were infected with typhus and transmitted the disease to others. Nicolle found that rats and mice uniformly displayed inapparent infection with typhus, even though they passed the pathogen to other rats and mice, and could pass the pathogen back to the guinea pig where symptoms were induced again. Nicolle extrapolated the phenomenon to other diseases caused by a variety of pathogens, and other researchers found that humans could also experience inapparent infection of some epidemic diseases.

Critical concepts underlying the dynamics of disease outbreaks were thus born: some people could spread an infectious disease far and wide even if they themselves did not get sick, and animals could serve as reservoirs for disease outbreaks. Inapparent infection in humans explained how typhus persisted in nature and broke out in seasonal waves.[195] "The new concept of inapparent infections that I introduced to pathology," Nicolle wrote, "is, without a doubt, the most important of the discoveries that I was able to make."[195] Nicolle also discovered that young children played a central role in typhus epidemics; they contracted only a benign form of typhus, sometimes completely inapparent, and hence served as a reservoir for lice to spread the disease.

Shortly after Nicolle discovered that lice are vectors of typhus, a tragic opportunity arose to apply his discovery on a massive scale. During World War I, typhus ripped through the Serbian, Austrian, and Russian forces along the Balkan and Eastern Fronts, reaching a mortality rate of 70 percent at the peak of the epidemic.[40,193] Nearly all of the Serbian doctors contracted typhus, and about one-third of them died. But typhus did not make inroads on the Western Front, though trench fever, another lice-borne (yet nonfatal) disease, did.[40] Constant delousing efforts on the Western Front prevented typhus from gaining traction; "the mortality of lice in this war," wrote Zinsser shortly afterward, "must have been the greatest in the history of the world."[40]

Nicolle understood the full implications of his recent discovery. "If in 1914 we had been unaware of the mode of transmission of typhus," wrote Nicolle, "the war would not have ended by a bloody victory. It would have ended in an unparalleled catastrophe, the most terrible in human history. Soldiers at the front, reserves, prisoners, civilians, neutrals even, the whole of humanity would have collapsed."[195]

But the disease took charge where lice populations were unchecked. Immediately after the war, during the 1917–23 Russian Civil War between the Bolshevik Red Army and the White Army, typhus sickened 30 million people, of whom 3 million died.[40] Similarly, in World War II, typhus often ran rampant in the filthy and crowded Nazi concentration camps. Perhaps the best-known victim of the Nazis is Anne Frank. Unlike other members of the secret annex who were gassed or shot, Anne and her sister Margot were transferred from Auschwitz to Bergen-Belsen where first Margot, and then Anne, died of typhus in the winter of 1945.[201]

Nicolle's finding of lice transmission marked the conclusion of two decades of fast-paced discoveries in which insects and other arthropods were identified as the vectors of numerous epidemic diseases, including those that ruined civilizations, turned the tide of wars, forced human migrations, and facilitated the conquest of the Americas. Nicolle was recognized for his experimental identification of the typhus vector with the 1928 Nobel Prize in Physiology or Medicine.

A typhus-like disease expressed in people who were not lousy was discovered by Nathan Brill in 1898 in New York City.[189] The disease chiefly affected Jews who had emigrated from Eastern Europe. Zinsser hypothesized that the disease was a reemergence of typhus in people who had survived epidemic typhus earlier in their lives.[202] The *Rickettsia* remained dormant in the victim's tissues until it again initiated disease. The resurgent form became known as

Brill-Zinsser disease.[189] Zinsser hypothesized that previously infected humans can reinitiate a typhus epidemic long after they recover if lice feed on them when the disease reemerges.[202] This pathway to typhus outbreak was confirmed in 1958 in Yugoslavia.[189] "Typhus is not dead," wrote Zinsser. "It will live on for centuries, and it will continue to break into the open whenever human stupidity and brutality give it a chance."[40]

Typhus belongs to a family of closely related *Rickettsia* diseases.[192] Some, like typhus and the rodent form of the disease known as murine typhus, are transmitted by lice or fleas. Other *Rickettsia* diseases are transmitted by ticks and mites. The discovery of the murine form of typhus in the 1930s provided another target in the war against typhus.

Murine typhus hides in animal reservoirs, especially rats. The rat louse and rat flea serve as vectors of murine typhus between rats, and the rat flea can also transmit murine typhus from rats to people.[40] Although the rat flea prefers rats, it will shift its attention and transfer its murine typhus to people when its rat host dies. Rats, therefore, play a vital role in outbreaks of typhus and other horrid diseases, cementing their place at the center of human misery. Rats consume a great deal of human food, and have at times caused famine when their numbers were particularly high, such as in 1615 in Bermuda, 1878 in Brazil, and 1881 in India. Starving people cannot mount an effective immune defense against diseases such as typhus, so the rat can initiate disease from all sides.

Some people in the ancient world suspected that rats carried evils. Although epidemics typically were blamed incorrectly on a variety of forces, some natural, such as volcanic eruptions, earthquakes, or eclipses, and some imagined, such as Jewish conspiracies, throughout history some perceptive observers faulted rats and mice. The association between these rodents and disease may explain why ancient Jews classified all mouse species as unclean, why the followers of Zoroaster killed rats to serve God, why the Greek god Apollo Smintheus protected against disease and killed mice, why early Catholic worshipers invoked Saint Gertrude to save them from plague and mice, and why, in the fifteenth century, the Jews of Frankfurt were forced to pay a tax each year of five thousand rat tails.[40] But the greatest misery that rats conveyed was not typhus— rather, it was an unparalleled pandemic that destroyed civilizations of the Old World, twice.

[5]

Black Death

(541–1922)

> There was a deadly panic throughout all the city; the hand of God was
> very heavy there. And the men that died not were smitten with swell-
> ings: and the cry of the city went up to heaven.—1 Samuel 5:11–12[203]

The Oriental rat flea and the black rat instigated the Dark Ages. Together, they conveyed the most destructive disease the world has
ever known, bubonic plague. "Plague" derives from the Latin *plaga*,
meaning "blow," "stroke," "injury," or "misfortune." The plague, also
known as the Black Death, disintegrated the Roman empire in the
sixth century during the reign of its last great emperor, Justinian.[204]
Simultaneously, it initiated the collapse of the Persian empire.

Justinian's plague of 541–42 CE killed between 25 million and
100 million people in Europe and the Middle East.[115,204] The armies
of Muhammad found the previously impenetrable forces of the Romans and Persians suddenly vulnerable. The Roman empire devolved
into medieval nation-states, and European civilization spiraled down-
ward into what Francesco Petrarch, the father of humanism and the
writer of some of the earliest literature in Italian, referred to as the
time of darkness,[205] or the Dark Ages. Petrarch defined the time
of darkness as the period between the fall of the Roman empire

(coinciding with the first plague pandemic) and his own time in the fourteenth century (corresponding to the second plague pandemic, or the Great Mortality).

While Justinian's plague in the sixth century drove Europe into the Dark Ages, the Great Mortality of the fourteenth century coincided with the Renaissance. Petrarch glorified the culture of the Roman empire, and felt that history subsequent to its fall was not worth recounting. He summarized this feeling with the statement, "What else, then, is all history, if not the praise of Rome?"[205] Nevertheless, during Petrarch's lifetime, everything fell apart again, and he recorded the events of the plague pandemic for posterity.

The plague pandemic of 1347–52 originated around 1334 in the Mongol empire and overwhelmed Asian nations.[206] Hidden in the fur of black rats that stowed away in the caravans of traders and soldiers, fleas carried the plague from Central Asia along the Silk Road to the Byzantine capital of Constantinople at the end of 1347.[207] An eminent scholar in Constantinople wrote that the disease "did not spare those of any age or fortune. Several homes were emptied of all their inhabitants in one day or sometimes in two. No one could help anyone else, not even the neighbors, or the family, or blood relations."[208]

Bubonic plague played a curious role in the annals of combat, as it was the agent of the earliest known germ warfare. Caffa, an important and impenetrable trade city on the east coast of the Crimea, was the object of intense battles between different Mongol khans, allied with Genoa, on the one hand, and Venice, on the other. In 1344, war broke out anew between the Genoese merchants in control of Caffa and a Mongol army. A Genoese resident, Gabriel de Mussis, documented the subsequent siege. He reported that three years into the siege, "the Tartars, fatigued by such a plague and pestiferous disease, stupefied and amazed, observing themselves dying without hope of health ordered cadavers placed on their hurling machines and thrown into the city of Caffa, so that by means of these intolerable passengers the defenders died widely. Thus there were projected mountains of dead, nor could the Christians hide or flee, or be freed from such disaster . . . And soon all the air was infected and the water poisoned, corrupt and putrified."[209]

The Genoese, some infected with plague, fled Caffa in their ships to ports in Sicily, Sardinia, Corsica, and Genoa, and hence facilitated the rapid spread of the pandemic.[209] De Mussis lamented his inadvertent role in spreading the plague. "We Genoese and Venetian travelers," he wrote, "of whom scarcely 10 survived of 1,000, as though accompanied by malign spirit, entered our homes . . . relatives and friends and neighbors came to us from all sides. . . . woe to us who carried the darts of death, as they held us with embraces and kisses while we spoke, from our mouths we were compelled to pour out poison with our words. Thus returning to their own homes they soon poisoned the entire family."[209] As the deaths mounted beyond society's ability to cope, "the great and the noble were hurled into the same grave with the vile and the abject, because the dead were all alike."

From Constantinople, the disease spread with lightning speed throughout the Middle East and Europe between 1348 and 1350, as rats scurried onto wagons and scampered onto ships with packages of textiles and food. A chronicler in Aleppo, shortly before his own death from the plague in 1349, wrote that "the plague did its work like a silkworm . . . It destroyed mankind with its pustules . . . How amazingly does it pursue the people of each house! One of them spits blood, and everyone in the household is certain of death. It brings the entire family to their graves after two or three nights."[210] A writer in Florence described that bodies and soil were layered in mass graves in the same manner that "one layers lasagna with cheese."[207] By 1352, half of Europe's population had perished, and it would take more than one hundred fifty years for the population to recover.[115,207] The norms of European society collapsed, and a new social order took hold.

Even observers found it difficult to believe the scale of the disaster. Petrarch wrote, "Will posterity believe these things, when we who have seen it can scarcely believe it, thinking it a dream except that we are awake and see these things with our own open eyes, and when we know that what we bemoan is absolutely true, as in a city fully lit by the torches of its funerals we head for home, finding our longed-for security in its emptiness? O happy people of the next generation, who will not know these miseries and most probably will reckon our testimony as a fable!"[211]

European reaction to the pandemic accelerated the spread of the pathogen. The flagellant movement, in which people chastised themselves with scourges, gained widespread popularity, as people believed that extraordinary atonement was required to appease God's wrath, and these masochists walked from town to town seeking converts and spreading the disease in their wake.[207] Plague doctors, outfitted in a birdlike mask and aromatic herbs to stop poisonous fumes, red glasses to ward off evil, and a long waxed overcoat to block bodily fluids, transmitted fleas from house to house.[115] Many people embraced hedonism, in the belief that the end of the world was at hand, and so neglected healthy habits and ignored societal norms. "People behaved as though their days were numbered," wrote an observer, "and treated their belongings and their own persons with equal abandon."[212] Considering each day to be their last, people failed to till the land or tend their flocks, and subsequent malnutrition weakened their immunity. Even bonds between parents and their children unraveled. "Fathers and mothers," according to a typical description, "refused to nurse and assist their own children, as though they did not belong to them."[212]

It was widely believed, as one scholar wrote, that the plague originated "from wicked men, children of the devil, who with venoms and diverse poisons corrupt the foodstuffs with evil skill and malevolent industry."[213] Christians, including the flagellants, accused Jews of poisoning wells to initiate the plague, and then burned them alive in retribution for their imagined culpability while destroying about a hundred Jewish communities, primarily in Germany but also scattered throughout Europe.[207,214-19] Some interrogators stated that the Jews obtained their poison from the basilisk, a mythical creature that could turn men into stone.[216]

One prominent scientist of the day, Konrad of Megenberg, concluded that Jews were probably not to blame: "But although the

Jewish people are justly detested by us Christians in accordance with the fundamentals of the Catholic faith, . . . nevertheless it does not seem to me that the said opinion concerning the cause of so general a mortality throughout the whole world, with all due respect to whomever is expressing it, can be totally and sufficiently maintained. My reasoning is as follows: It is well known that in most places where the Hebrew people had remained, they themselves had died in droves from the same exact cause of this common mortality . . . But it is not likely that the same people who ardently desire to multiply themselves upon the land should with malice aforethought destroy themselves and others of the same faith . . . Moreover, even after all the Jews in many places have been killed and completely driven out for nearly two years prior, the Death now first strikes these same places with a strong hand and powerfully conquers the men who remain there."[218] Pope Clement VI agreed, and issued a protective mandate despite his admission that "we justly detest the perfidy of the Jews."[219]

All manner of other possible causes were invoked, in addition to well poisoning by Jews. Many people blamed the wrath of God. The flagellants preached a sermon in which God says, "In a few years, much misery happened: earthquakes, hunger, fever, locusts, rats, mice, vermin, pocks, frost, thunder, lightning, and much disorder. I sent you all this because you haven't observed my holy Sunday."[220]

Despite this widely held belief that God had released the plague in response to man's many sins, it was difficult to understand God's timing. Petrarch wrote, "I do not deny that we deserve these misfortunes and even worse; but our forebears deserved them too, and may posterity not deserve them in turn. Therefore why is it, most Just of judges, why is it that the seething rage of Your vengeance has fallen so particularly hard upon our times? Why is it that in times when guilt was not lacking, the lessons of punishment were withheld? While all have sinned alike, we alone bear the lash."[211] In comparison to the punishment at hand, Petrarch thought that the wrath of God at the time of Noah "would have been a delight, a joke, and a respite."

While many attributed the plague to God's wrath, others found cause in astrological phenomena. The most prominent scientific treatise produced during the pandemic, by the Medical Faculty of the University of Paris, noted that "the distant and first cause of this pestilence was and is a certain configuration of the heavens. In the year of our Lord 1345, at precisely one hour past noon on the twentieth day of the month of March, there was a major conjunction of three higher planets in Aquarius . . . For at that time, Jupiter, being hot and wet, drew up evil vapors from the earth, but Mars, since it is immoderately hot and dry, then ignited the risen vapors, and therefore there were many lightning flashes, sparks, and pestiferous vapors and fires throughout the atmosphere."[221]

A musician in the papal court at Avignon, like many others, attributed the Black Death to atmospheric causes and invoked biblical imagery: "On the first day it rained frogs, serpents, lizards, scorpions, and many venomous beasts of that sort. On the second day thunder was heard, and lightning flashes mixed with hailstones of marvelous size fell upon the land, which killed almost all men, from the greatest to the least. On the third day there fell fire together with stinking smoke from the heavens, which consumed all the rest of men and beasts, and burned up all the cities and castles of those parts."[222] Such wreckage was due to infection carried by the "fetid breath of the wind blowing southwards from the plague regions." The overwhelming devastation also led to fanciful exaggeration, such as the report that in a Greek territory "men and women and every living animal turned into the likeness of marble statues."[223]

Scholars more disposed to logic searched for earthbound, naturalistic explanations. One prominent cleric argued that earthquakes allowed noxious fumes to escape into the atmosphere.[207] It was a reasonable hypothesis, because the onset of the plague coincided with earthquake activity in parts of Europe. The Medical Faculty of the University of Paris agreed that earthquakes were partly to blame, and also argued that disruption of the seasons and shooting stars played a role.[221] Other scholars, approaching much closer to the truth, blamed contagious passage of the poison from person to person.[224] Pope Clement VI apparently agreed since he bolted from his plague-beleaguered city of Avignon.[207] The Muslim world could not accept the theory of contagion because it implied that the disease spread outside the control of God. A prominent Muslim scholar, based on his own observations, concluded that the plague was contagious,[225] and for this piece of scholarship he was lynched.[207]

Some perceptive observers noted an association between the Black Death and rats, and even chronicled efforts to destroy rats. One, before his death from the disease in 1348, wrote that "it rained an immeasurable quantity of vermin, some as big as eight hands, all black and with tails, some alive and some dead. This frightening scene was made worse by the stench that they emitted, and those who fought against the vermin fell victim to their venom."[223]

Other intellectuals noticed a relationship between infection and the handling of the personal belongings of the sick. One scholar wrote of the Black Death that "not only did it infect healthy persons who conversed or had any dealings with the sick, making them ill or visiting an equally horrible death upon them, but it also seemed to transfer the sickness to anyone touching the clothes or other objects which had been handled or used by victims."[212]

Medical providers recommended a diversity of preventive measures, such as the consumption of select wines and the burning of fragrant plants.[226] Many foods were to be avoided. For example, "slimy fishes such as lampreys and eels and rapacious fishes, such as dolphin, shark, tunny-fish, and similar fishes should be forbidden."[213] Those at risk of infection included people "that live by a bad regimen, indulging in too much exercise, sex, and bathing; those who are weak and thin and very fearful. Also infants, women, and the young, and those whose bodies are fat and have a ruddy complexion."[221] Imagination and fear of death must also be avoided: "This influence is of such great force that it will change the form and figure of the infant in the mother's womb."[213] Sadness was thought to contribute to the plague, though not in equal measure for everyone, for "it strikes intellectuals the hardest, least of all idiots and indolents."[227] Doctors employed bloodletting and recorded it to be most effective in the middle of the third quarter of the moon, "provided that the moon in such times is not seen in a sign unfavorable for bloodletting, such as Gemini, Leo, Virgo, Capricorn, and some others."[213]

Ignorance of the cause inevitably led to a failure of prevention and treatment. A prominent Italian scholar pointed this out at the height of the plague in Florence: "Against these maladies, it seemed that all the advice of physicians and all the power of medicine were profitless and unavailing. Perhaps the nature of the illness was such that it allowed no remedy: or perhaps those people who were treating the illness (whose numbers had increased enormously because the ranks of the qualified were invaded by people, both men and women, who had never received any training in medicine), being ignorant of its causes, were not prescribing the appropriate cure."[212] Another observer wrote of the pitiful state of medical care, noting that "doctors from every part of the world had no good remedy or effective cure, neither through natural philosophy, medicine, or the art of astrology. To gain money some went visiting and dispensing their remedies, but these only demonstrated through their patients' death that their art was nonsense and false."[207] A chronicler in Siena wrote that "the more medicine people are given the quicker they die."[224] Petrarch noted the additional stress induced by "not knowing the causes and origin of the evil. For neither ignorance nor even the plague itself is more hateful than the nonsense and tall tales of certain men, who profess to know everything but in fact know nothing."[211]

The overwhelming mortality led to sadness and despair. Petrarch bemoaned, "Where are our sweet friends now? . . . We should make new friends, but where or with whom, when the human race is nearly extinct, and it is predicted that the end of the world is soon at hand? We are—why pretend?—truly alone."[211]

The mortality led to bizarre behavior, including a dancing mania where dancers fell to the ground so that onlookers could trample them to effect a cure.[207] The chroniclers of such dancers may have been describing St. Vitus's dance, a neurological symptom of bubonic plague.

Artists illustrated some of the stranger conduct, such as the "Dance of Death." Sometimes, Death was portrayed as a chess player, conveying the arbitrary behavior of the plague.[207] A macabre fashion developed among the wealthy to design their tombs before their death and to employ revolting imagery. A typical tomb of this style had worms emerging from and slithering into the arms and legs of a stone replica of the deceased, and frogs sitting upon the eyes, lips and genitals.

The plague also crept into a wide variety of literature, the most heart-rending of which may have been Shakespeare's *The Most Excellent and Lamentable Tragedie of Romeo and Juliet*.[228] Romeo did not receive Friar Laurence's letter describing his ruse because his messenger was locked up to prevent the spread of plague; as a consequence, Romeo drank the poison next to the apparently dead Juliet in the Capulet crypt.

After 1352, the pandemic subsided. But then the plague returned in more than one hundred localized epidemics; outbreaks occurred somewhere in Europe every generation until the late 1700s. By then, the brown rat had outcompeted and replaced the black rat in most of Europe, as changes to human society favored the brown over the black. These changes included the widespread construction of water drainage systems and the separation of granaries and stables from homes.[229] Since the brown rat preferentially inhabited sewers rather than homes, contact with people fell along with the risk of plague. Plagues continued to devastate regions of Africa and Asia where the black rat dominated.[206]

Yersinia pestis (1894)

> A man seemingly convalescent takes his food, refers to it with a cheerful smack of the lips, tumbles back, and dies in a few minutes; another, with no fever for days, and feeling his ordinary self, walks in the verandah and drops down dead.—**Dr. James Cantlie**, Hong Kong, 1894[230]

After a considerable hiatus, plague returned to the war-ravaged Yunnan province of China in the 1860s.[115] It moved east to coastal China and then hit Hong Kong in 1894. The rate of infection varied by social class, and hence by economic standing. A Hong Kong surgeon noted, "Of the various races in Hong Kong the following is the order of susceptibility—Chinese, Japanese, Hindus (from India), Malays, Jews, Parsees, and English."[230] When a patient fell under the care of a European doctor, the surgeon reported, the chance of survival was 20 percent; when receiving Chinese treatment, the survival rate was 3 percent. One hundred thousand Chinese residents fled the city, most heading to Canton, which itself had by that time one hundred thousand dead residents out of a population of 1 million.[231] As a consequence, wrote one correspondent, "the busy, crowded thoroughfares [of Hong Kong] are but the skeleton of their former selves, and a sad stillness prevails."[231]

Finally, an outbreak of plague corresponded to a period with the intellectual and technical capacity to study it. The man on the ground was the Scottish physician James Alfred Lowson. Only twenty-eight years old, Lowson superintended Hong Kong's Civil Hospital, where he made the first plague diagnosis on May 8, 1894.[232]

Lowson discovered that a crowded Chinese neighborhood harbored an outbreak of plague, and he recommended a series of public health measures to Hong Kong's Sanitary Board: thorough searches for plague victims, disinfection of their homes, quick removal of the dead, and quarantine of patients in special hospitals.[233] Hospitals were quickly established, including a hospital ship run by European doctors for European patients and a hospital on land run by Chinese doctors for Chinese patients.[234] Additional facilities to handle the deluge of plague patients swiftly sprang to life, though their offerings were meager, with no beds, blankets, or mosquito nets available. Indian and Japanese patients were provided with mattresses, but not Chinese patients, due to the rationing of limited supplies on the basis of race.

A prominent Chinese board member and physician, Ho Kai, who along with Patrick Manson and others had founded the Hong Kong College of Medicine, warned that home searches, disinfections, removal of the dead, and quarantine actions would rile the Chinese residents.[233] Ho Kai's intuition was spot on. "As usual with all

Asiatics," read an editorial in the *British Medical Journal*, "foreign interference in medical treatment was resisted . . . The native Chinese doctors are ever on the watch to find opportunities to cast a slur and heap ignominy on the heads of English doctors, and the present epidemic was seized on by them and by all anti-English as a golden opportunity of gratifying their heart's desire."[231] Pamphlets distributed among the Chinese stated that English medicines were prepared from the bodies of the previous day's dead, or that English doctors gave the patients brandy and then crushed them with ice. When health teams attempted to remove plague patients from a traditional Chinese hospital, riots broke out.[233] Physicians were stoned,[231] and henceforth carried on their duties with pistols on their belts, while the British military reestablished order.[233]

The Chinese continued to frustrate British medical authorities. A British medical editorial commented, "They will not report the presence of plague in their houses; they object, even to the point of positive resistance, to the intrusion of sanitary authorities; they attempt to smuggle their sick from the colony by all kinds of subterfuges; in fact, they do all in their power to hide the disease from the Government. This is partly accounted for by the dread of being submitted to foreign medical treatment, by the knowledge that if a case of plague is found in a house all the other inmates have to turn out; but mostly by the inborn hatred of the Chinaman to have the seclusion of his home and the 'feng-shui' of his establishment disturbed."[235] A British authority in Hong Kong wrote of the Chinese after the epidemic, "Educated to insanitary habits, and accustomed from infancy to herd together, they were unable to grasp the necessity of segregation; they were quite content to die like sheep, spreading disease around them, so long as they were left undisturbed, and they preferred to see their sick friends and relatives suffering unspeakable miseries rather than be parties to their removal to European hospitals, where every comfort was provided."[236]

Lowson, accompanied by soldiers, trod from house to house in search of plague victims. They were not hard to find. Lowson wrote, "On a miserable sodden matting soaked with abominations there were four forms stretched out. One was dead, the tongue black and protruding. The next had the muscular twitching and semi-comatose condition heralding dissolution. In searching for a bubo we found a huge mass of glands . . . Another sufferer, a female child about ten years old, lay in accumulated filth of apparently two or three days. The fourth was wildly delirious."[237] Soon, the health teams began to encounter one hundred new cases per day.[233] Most plague victims had lost their faculties, but those who were still conscious reportedly "prayed for death" and "beat themselves on the floor to hasten the end."[231] One district of Hong Kong was so infested with bubonic plague that the authorities condemned it, expelled the residents, and walled up its streets with brick and mortar.

Doctors noticed that the plague did not appear to spread via direct contagion from person to person, but was associated with filth.[230] Nurses attending to the ill did not get sick, though some of the soldiers who cleaned the accumulated grime of houses of the dead did. Members of the many inspection teams did not contract the plague, but they also did not have intimate contact with filth. These outcomes were consistent with the Chinese belief that the plague toxin originated in the ground and rose up to infect those closest to the ground first; hence, the Chinese observed that the plague struck in sequence rats, poultry, goats, sheep, cows, buffalo, and humans. "Man, being the tallest in point of distance of the head from the earth, is that last to be seized."[230] The filth hypothesis also struck a chord with European doctors. "The word 'soil' comes, not only figuratively, but actually, near the mark," commented the *British Medical Journal*, "for a Chinaman of the coolie class seldom or never cleans his abode, but allows the refuse of the household to litter the floor, where it gets packed and trodden down into a veritable midden."[235] Once the heat of summer arrived, doctors thought, the germs that had lain dormant in the midden came back to life.

Competing scientists arrived in Hong Kong to search for the responsible germ. The Japanese imperial government sent a team led by Shibasaburo Kitasato and his rival Tanemichi Aoyama.[238] Both had trained under Koch in Berlin after completing their medical studies in Japan. Kitasato was one of the world's most famous bacteriologists for his isolation in 1887 of the tetanus bacillus, and in 1892 he became the first foreigner given the title of professor by the German government.[234,239] The Japanese team arrived in Hong Kong on June 12, 1894.[240] Three days later, Alexandre Yersin arrived alone from French Indochina (comprising Laos, Cambodia, and Vietnam), dispatched there by the French colonial minister upon the suggestion and under the auspices of Koch's rival, Pasteur.[233,241]

Yersin was born in Switzerland, but trained under Pasteur and became a nationalized Frenchman before joining the French colonial medical corps.[241] He also briefly trained under Koch.[242] During an autopsy of a rabies patient in Paris, Yersin's knife slipped while cutting the spinal cord and drew blood from his finger. Pasteur instructed his assistant, Emile Roux, to vaccinate Yersin against rabies with his newly introduced vaccine. So began the friendship among Yersin, Pasteur, and Roux.[241]

Yersin became known for his 1888 discovery with Roux of the diphtheria toxin, the first discovery of a poison produced by bacteria.[234] Kitasato and a colleague in Germany then showed that animals produce an "anti-poison," or antitoxin, when injected with a toxin.[243] These antitoxins later became known as antibodies. Roux extended Kitasato's discovery to produce antitoxin to diphtheria in horses, which he used to save the lives of children infected with diphtheria—thus inventing serotherapy.

Inspired by the Scottish explorer David Livingstone, Yersin then mounted expeditions of French Indochina to map the incidence of endemic diseases, such as malaria and smallpox, with the goal of protecting French colonists.[244] He was the first European to explore Vietnam's central highlands and, therefore, the first European seen by the mountain people who lived there.[242] During his explorations he contracted severe attacks of malaria and dysentery.

Despite these achievements on behalf of France, Yersin did not have the stature of Kitasato when he arrived in Hong Kong.[233] Furthermore, the race between Kitasato and Yersin to identify the most dangerous germ in human history continued the old rivalry between Koch and Pasteur.

Kitasato had a three-day head start. He also had the assistance of Lowson, who gave him ready access to plague patients. On June 14, Kitasato discovered a bacillus during Aoyama's autopsy of a plague victim.[240] The autopsies were highly risky; Aoyama and an assistant contracted plague, which prevented Aoyama from competing with Kitasato in the contest to find the pathogen.[238] Aoyama survived.[234]

Kitasato found that the bacillus was present in the blood, lungs, liver, spleen, and the swollen lymph nodes in the groin (the characteristic buboes of bubonic plague).[240] The patient in question had been dead for eleven hours, so Kitasato was not sure that what he had found was the responsible bacillus. He inoculated a mouse with a piece of the patient's spleen, and with various other tissues he inoculated mice, guinea pigs, rabbits, and pigeons. The mice were dead in two days and contained the same bacillus. The guinea pigs and rabbits also died in short order, and also contained the suspected bacillus. Then he found what appeared to be the same bacillus in other plague victims, in their buboes, spleen, lungs, liver, blood, brain, and intestines.

Lowson felt sure that Kitasato had discovered the responsible germ. He wired details of the finding to the *Lancet* on June 15, the day after Kitasato first found the bacillus and the same day that Yersin arrived in Hong Kong.[233] A week later the *Lancet* announced that Kitasato had "succeeded in discovering the bacillus of the plague."[245]

Lowson reserved all of the plague victims' bodies for Kitasato, and hence stymied Yersin's progress.[233] Lowson may have been motivated to bar Yersin from the plague morgue by colonial rivalries between Britain and France, or by prestige (Kitasato was far more famous than Yersin), prudence (he was convinced that Kitasato had already found the bacillus), or scientific jealousy. Lowson wished to investigate the cause of plague himself, and he had made some attempts at this with rabbits and guinea pigs, but his duties left him insufficient time for meaningful research. "Our time," he wrote, "being

Fig. 5.1. Shibasaburo Kitasato in Koch's research institute, 1889. When Koch died, Kitasato built a shrine in the institute, and each year that he worked there, upon the anniversary of Koch's death, Kitasato performed a Shinto ceremony for the memory of the departed soul.[239]

entirely taken up by practical work in connection with the treatment of the plague—for which no fame is secured—we had so little time to look to the more purely scientific side of the question."[237]

Yersin met Kitasato in the midst of an autopsy.[234] They had a stilted conversation in their only common language, German, of which Yersin was a poor speaker. Yersin was amazed to see that while the Japanese scientists carefully inspected the blood and internal organs, they did not focus on the buboes. Yersin felt that he must gain access to the buboes of cadavers, but he could not see how to do it until an Italian missionary who served as his guide recommended an unorthodox approach.[233,241] Five days after his arrival, on June 20, Yersin bribed the English sailors who disposed of the corpses to gain access to them. "A few dollars conveniently distributed," wrote Yersin in his diary, "and the promise of a good tip for every case have a striking effect."[234]

While a body lay in its coffin in a bed of lime, Yersin exposed a bubo, excised it, and dashed off toward his laboratory, all in under a minute. Yersin's rudimentary laboratory was at first an open porch, and then a straw hut.[234] Upon viewing the sample through his microscope, he saw *"une véritable purée de microbes."*[246] Yersin wrote in his lab notebook that "this is without question the microbe of plague."[241] He named the pathogen in honor of his mentor, *Pasteurella pestis*[247] (*pestis* is the Latin word for "curse" or "bane").

Yersin grew the microbes extracted from the cadavers' buboes and injected them into mice and guinea pigs, which died the next day; the rodents also displayed typical buboes of plague.[234] He found the same microbes in the rodents' lymph nodes, and he also found them in dead rats in Hong Kong.[241] Yersin used this evidence to successfully appeal to officials for access to cadavers.

The *Lancet*, as a British publication, displayed prejudice against Yersin. Armed with evidence provided by Lowson,[233] the *Lancet* published editorials on August 4 that both reiterated the discovery of Kitasato's bacillus and warned readers that "there will probably be many local *savants* keen on discovering a specific bacillus, and it is necessary to caution the profession against accepting all the statements which will no doubt be put forward in this respect."[232] In another editorial, the *Lancet* wrote that Yersin had "discovered another bacillus, which he, too, claims to be the essential cause of the disease; and others, equally anxious to discover something, have swelled the list, so that, as our correspondent says, 'the varieties of plague bacilli now outnumber the leaves in Vallombrosa.' Amongst all these claimants for priority it will be difficult to decide who of them, if any, is entitled to have his fame handed to posterity as the discoverer of the cause of plague; but, as we have before remarked, the name of Professor Kitasato is a guarantee of accuracy in observation and care in research, and when the opportunity is given for the review of his work it will probably be found to meet the severest tests."[248]

Later that month, the *Lancet* and the *British Medical Journal* published slides of Kitasato's pathogen, which had been provided

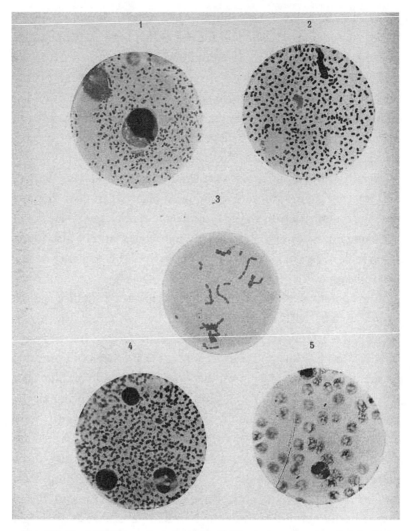

Fig. 5.2. Yersin's microscopic images of (1) a bubo of a plague-infected Chinese patient; (2) a lymph node from a dead rat infected with plague; (3) a culture of plague bacteria in bouillon; (4) a lymph node of a mouse inoculated with plague; and (5) blood collected from a plague victim fifteen minutes after his death with two bacteria present in the image.[246]

by Kitasato and Lowson.[249,250] The forms of the bacillus were unusually diverse. The *British Medical Journal* also reported on Yersin's discovery of a bacillus and the quick mortality it induced in rats, mice, and guinea pigs whether introduced by inoculation or by feeding the animals on the tissues of plague victims.[250] The *Lancet* thought that Yersin must be mistaken, since "Professor Kitasato is such an accurate and reliable observer that we cannot conceive that he has rushed into print without having first satisfied himself as to the accuracy of his observations and experiments."[251]

But Kitasato had, in fact, rushed into print.[233,234] He had discovered his bacillus just two days after his arrival in Hong Kong and six days before Yersin's discovery, and he undoubtedly felt the pressure of the competition. Despite his normally cautious approach to science, Kitasato also felt pressure coming from Lowson, who wanted Kitasato to get the honor of discovery before Yersin.[233] Kitasato's bacterial cultures were contaminated, which gave rise to the diverse morphology seen in the British medical journals. Initially, Kitasato could not determine whether his bacillus stained gram positive or negative[240] (such a determination was standard practice in bacteriology), and he published conflicting accounts of his findings.[234] He then found that his bacillus stained gram positive. Kitasato's coworker and rival Aoyama criticized him for this muddled work.[238]

Yersin, however, was able to determine that the bacillus could not be stained using Gram's method (it was gram negative).[241] The two researchers had, in fact, documented different bacterial species.[206,252] Nevertheless, reports around the world referred to "the bacillus of Kitasato-Yersin."[252] Although for decades both Kitasato and Yersin were honored for their roles in the discovery of the pathogen[239,253] (Lowson was not), eventually the credit for the discovery went to Yersin. In 1970, twenty-seven years after Yersin's death, the pathogen was renamed *Yersinia pestis*.[233,247]

Up until his death in 1935, Lowson maintained his belief that Kitasato and not Yersin had discovered the plague pathogen.[233] Kitasato, however, did not persist in ascribing his pathogen to the disease once he had the chance to investigate another outbreak in

Kobe, Japan, in 1899. "In Kobe," he wrote, "I had the opportunity of examining plague patients, and in each case I recognized the truth that the bacillus of Yersin was the specific one."[252] The chief medical officer of the Japanese Navy lamented the turn of events. "Now the honour of the discovery of the plague bacillus must belong to Yersin alone," he wrote, "and we much regret that so distinguished a bacteriologist as Kitasato should have made so incredible an error in his search for the microbe."[252] But then considerably later, in the 1920s, Kitasato wrote that Japanese scientists (presumably himself) had discovered the plague bacillus.[234,254]

As with so many other contagious diseases, there were those who attempted to self-inoculate with bubonic plague. The earliest of these attempts appears to have been that of an English army doctor in Egypt, who, in 1802, wiped the contents of a bubo over his groin and injected it into his arm.[206] He then died of plague.

The Hong Kong outbreak led to a more fruitful preventive approach. When Yersin discovered the plague bacillus in Hong Kong, he immediately mailed a sample to the Pasteur Institute in Paris—he sent the live plague culture in a sealed glass tube held within another tube packaged inside a piece of bamboo.[242] Fortunately, the package reached its destination. Roux cultured the sample and prepared an antiplague serum from it.[247] Yersin tested the new treatment in China in 1896 with some success.

In 1897 Yersin was transferred to French Indochina, where he established the Pasteur Institute in Nha Trang, and then founded and led the Medical School of Hanoi.[238] Yersin also established an agricultural station for the cultivation of Brazilian rubber trees, constructed a telescope within a dome upon the roof of his Vietnamese home, and in 1915 founded a new Vietnamese agricultural station to cultivate cinchona trees for the production of quinine to treat malaria.[255] This last effort, he felt, was critical to secure French supplies of quinine during World War I.[242] In 1940, at the start of the next world war, Yersin visited Paris and then caught the last flight out before the German invasion. He died three years later in Nha Trang. Yersin's tomb in Suoi Dau includes a pagoda for religious rites, and his epitaph reads, "Benefactor and humanist, venerated by the Vietnamese people."[256] The villagers in Nha Trang, where Yersin lived for half a century, called him "Monsieur Nam," meaning "Mr. Fifth," a reference to the five stripes of his French uniform designating him as a colonel, and also a way to avoid having to pronounce his foreign name.[242]

Rat Fleas (1897–1922)

That day, 2 June 1898, I felt an emotion that was inexpressible in the face of the thought that I had uncovered a secret that had tortured men since the appearance of plague in the world.—**Paul-Louis Simond**, 1898[247,257]

Although Yersin identified the responsible pathogen and its infectivity of both rats and humans, he did not discover how the pathogen infected its hosts, though he suspected that rats were principally to blame.[258] Yersin also hypothesized that the pathogen could spread through exposure to infected soil.[259,260] Kitasato speculated that the bacillus could infect humans through respiration, by entering through a wound, or via the intestinal tract.[206,240] He also wondered whether flies and other insects as well as rats and mice might propagate the disease.[206] It is curious that neither Yersin nor Kitasato made the connection to rat fleas, but their work in Hong Kong occurred in 1894, just before the critical discoveries of insect vectors of malaria and yellow fever.

A Hong Kong surgeon, James Cantlie, noted the preponderance of dead rats, but although he came close, he also did not connect the disease to rat fleas. "The rats seem to leave their haunts in sewers and drains," Cantlie wrote, "and, escaping by their holes, seek the houses. They seem careless of the presence of men, and run about in a dazed way with a peculiar 'spasm' of their hind legs. They are frequently found dead on the bedroom floor, but more often the foul odours from their dead bodies under the flooring is a proof of what is going on."[230] The Chinese also noted that rats died in exceptional numbers two or three weeks before the plague struck humans and regarded rats as "heralds of the coming evil."[261] "The superstitious Chinamen," noted a British medical report, "regarded these animals as the messengers of the devil, and endeavoured to drive them away."[262]

In one quarter of Canton, officials organized the collection and burial of twenty-two thousand dead rats,[261] and fifteen hundred rats were collected from one street in Hong Kong.[262] When inspected, the rats were found to harbor the plague bacillus, and Cantlie pointed out that their physical and behavioral symptoms were the same as those of humans.[230] The inoculation experiments that conveyed the plague bacillus from humans to rats and other rodents also suggested an important connection. "The rat infection therefore cannot be passed lightly by," wrote Cantlie, "but must enter seriously into the conditions affecting the spread of the disease. Even assuming that rats are affected, and we cannot doubt it, they may be merely affected in common with man, or they may be the actual carriers of the disease to human beings."[230]

The first written hypothesis blaming rat fleas came in 1897 from Masanori Ogata, who wrote that "one should pay attention to insects like fleas, for as the rat becomes cold after death they leave their host and may transmit the pest virus direct to man."[234] Ogata found dead plague rats and collected their fleas. He crushed them and injected them into two mice, one of which died of plague.[263] Ogata considered fleas to be one of many possible routes of plague infection.[247]

Independently, the French physician Paul-Louis Simond experimentally demonstrated the critical role of the rat flea. Simond entered the field of plague research during the high point of his productive career. He directed a leprosarium in French Guiana from 1882 to 1886, where he survived yellow fever.[264] He then worked throughout East Asia before taking a post in 1895 at the Pasteur Institute in Paris, where he studied protozoan parasites similar to malaria.[247] He discovered male sexual elements in these parasites, which was an important step in the investigation of the natural history of malaria. Patrick Manson dispatched Simond's paper describing these sexual elements to Ronald Ross in India; unfortunately, Ross did not recognize the significance of the discovery.

In 1897, Simond filled Yersin's vacant post in Bombay with the assignment to study the utility of the new antiplague serum.[265] The Indians' violent reaction to the antiplague serum had motivated the authorities to bar Ross's research on malarial patients, in one of Ross's many set-backs.[65] Late that year, Simond, overtaxed by the intensity of his work, fell ill with malaria.[264]

The following April, Simond was dispatched to Jurrachee (Karachi) to investigate an outbreak of plague.[265] Just as Yersin had experienced in Hong Kong, Simond found his research in Karachi thwarted by British authorities who barred his entrance into plague hospitals. Simond discovered, as had Yersin and Cantlie, sick and dead rats in the plague-stricken areas. In one house he found seventy-five dead rats.[247] Simond recorded a critical observation in his notes: "one day, in a wool factory, employees arriving in the morning noticed a large number of dead rats on the floor. Twenty laborers were ordered to clean the floor of the dead animals. Within 3 days, 10 of them developed plague, whereas none of the other employees became ill."[241] He visited a village where the preponderance of dead rats made the wise inhabitants reckon that plague was coming and hence flee to an isolated camp. "Two weeks later," wrote Simond, "a mother and daughter received permission to go back to the village to bring clothing from their house. They found several dead rats on the floor of the house. They picked up the rats by their tails and threw them out on the street and then returned to the camp. Two days later, both developed plague."[241]

"We have to assume," Simond wrote, "that there must be an intermediary between a dead rat and a human. This intermediary might be a flea."[241] Simond carefully watched the rats and observed that healthy individuals, which groomed themselves, sheltered few fleas, while sick rats were infested with them.[247] When the rats perished, their fleas departed from the cooling corpse in search of another rat or human, which meant that rats were particularly dangerous to handle immediately after their death.[266] Additionally, Simond noticed that some plague patients had small blisters on their skin that contained the plague bacillus, and he hypothesized that these were the sites of flea bites.[265]

Simond's rat flea hypothesis encountered the skepticism of other plague investigators.[265] But Simond was encouraged by his mentor, Laveran, who simultaneously supported Ross's seemingly crazy hypothesis that mosquitoes transmit malaria. Simond examined fleas taken from plague-ridden rats, and discovered a bounty of plague bacilli in their digestive tract, though fleas from healthy rats did not contain the bacilli. He then needed only to demonstrate the flea's role in infection from one rat to another.

Simond devised a simple experiment to test this idea. For the first stage of the experiment, he caught a sick rat in the home of a plague patient. It had several rat fleas scampering about its fur. He returned with the sick rat to his makeshift lab in the Hotel Reynolds in Karachi, where he placed the rat in a large glass jar with a mesh lid. Simond wanted to house additional fleas with the rat. "I took advantage of the generosity of a cat I found stalking the hotel premises," he wrote, "borrowing some fleas from it."[265] He added these fleas to the rat bottle. "After 24 hours," he wrote, "the animal I was experimenting on rolled up into a little ball, with its hair standing on end; it seemed to be in agony."

For the second stage of the experiment, Simond suspended a healthy rat in a wire mesh cage above the sick rat reposed at the bottom of the jar.[241] The two rats could not touch each other, but fleas on the sick rat were able to jump high enough to reach the healthy rat, and they thereby transmitted the plague (Simond had already discovered that rat fleas can jump up about 10 cm). The sick rat did not move from the bottom of the jar, and died the next morning. Simond found that its blood and organs were resplendent with Yersin's bacillus. The suspended rat died of plague after six days, its blood and organs also harboring an abundance of bacilli.[257]

Simond replicated the experiment with the same result. Additionally, when Simond caged sick and healthy rats together in a jar, but without fleas, the healthy rats did not contract the plague.[265]

Simond had performed the first experimental demonstration of a bacterial disease spread by insects.[247]

Other scientists rejected Simond's research, going so far as to proclaim it worthless, in part because they were unable to replicate his findings.[247,260,265] Investigators typically did not record which species of rat flea they used in their experiments, and this could account for the discrepant results.[260] However, in 1903, five years after Simond's critical study, investigators in Marseilles conducted confirmatory experiments.[267] Then, in 1906, an English commission studying plague in India also confirmed Simond's findings in a series of carefully controlled experiments with rats, guinea pigs, and monkeys.[260]

The English commission found that close contact between infected and healthy animals in the absence of rat fleas (including contact with pus from plague ulcers, urine, and feces) did not convey plague.[260] Similarly, offspring did not contract plague from suckling on infected mothers. Plague was also not transmitted through the air. However, if rat fleas were present, then the plague spread from animal to animal in proportion to the size of the flea population. This occurred whether or not animals had contact with infected soil. When guinea pigs were released into the homes of plague victims, the animals attracted rat fleas and contracted plague. The commission then tried to first eradicate fleas from plague homes using an acid solution of perchloride of mercury or the fumes of burning sulfur before releasing guinea pigs, but they were unable to kill all the fleas. The commission found that the blood of infected rats contained as many as 100 million bacilli per milliliter, making it inevitable that rat fleas sucking the blood of the infected rat would imbibe many bacilli. Taken together, the English commission's results finally led to scientific acceptance of the rat flea hypothesis.[265]

Meanwhile, other investigators claimed credit for the rat flea hypothesis and deflected recognition from Simond. In 1905, the *Indian Medical Gazette* published a paper by Captain W. Glen Liston of the British Indian Commission on rat fleas as the vector of plague, with language a great deal like that of Simond.[229] The following year, the *Indian Medical Gazette* referred to Captain Liston's rat flea theory as an epoch-making discovery.

Simond's findings explained why quarantines against plague were ineffective. The word "quarantine" comes from the Italian *quaranto giorni*—a period of "forty days" during which ships must isolate themselves before gaining permission to dock, unload cargo, and disembark passengers and crew.[115] Many port cities employed quarantines during the plague of 1347–52, turning away ships with sick people, and disallowing entry if people aboard a ship became ill during the quarantine. Yet plague entered these cities all the same, since rats were able to scamper down the mooring ropes of docked ships or swim the short distances to land.

Simond followed up his bubonic plague studies with three years running vaccination programs in Saigon, followed by five years of work on yellow fever in Brazil. In Brazil, he and his team confirmed the Reed commission results that the causal agent of yellow fever is a virus transmitted by *Aedes aegypti*.[268] Simond then replicated Gorgas's mosquito eradication program in Martinique to stamp out yellow fever on the island.

Simond hypothesized that a rat became infected with bubonic plague when, in the process of scratching a flea bite wound, it rubbed in flea feces that contained the bacillus.[269] He also speculated that infection could occur through a droplet of infected blood left behind by the flea.[257]

In 1914, the entomologist Arthur Bacot and the director of England's Lister Institute, Charles James Martin, discovered that the flea's feces were not normally to blame. At first blush, Bacot's invitation to join the plague project appeared strange. Bacot was not formally trained in science; he worked as a clerk, and he lacked experience working on fleas.[270] However, Bacot had a keen mind for entomology, and quickly discovered the details of the flea's life history. His successes led to a position as entomologist at the Lister Institute.

Bacot and Martin found that the flea's feces quickly dried up and contained few bacilli.[269] When the investigators deprived infected fleas of food for a day so that they did not defecate when fed upon

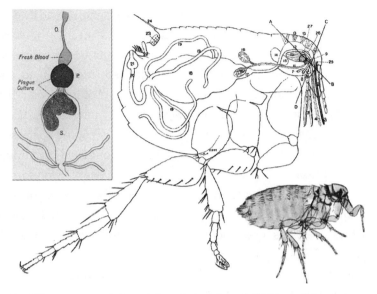

Fig. 5.3. Anatomy of the rat flea, *Pulex cheopis*.[260] The left inset shows the obstruction of the flea's proventriculus (P) by the plague culture and clotted blood; S = stomach, O = oesophagus.[269] The right inset shows a female of the common rat flea *Ceratophyllus fasciatus* chemically preserved and photographed during plague investigations in India in 1914.[272]

a rat, they still conveyed the bacillus to the rat. They found that the bacillus multiplies in the flea's stomach and proventriculus (similar to a gizzard) to create "a solid culture of plague," which, along with freshly clotted blood from the flea's last blood meal, blocks the passage of food, forcing the flea to seek constant nourishment. "The flea suffers from thirst," wrote Bacot and Martin, "and is persistent in its efforts to satisfy this appetite, but only succeeds in distending the oesophagus."[269] A single flea can harbor over 1 million bacilli.[271] Each time the ravenous flea bites into a host, whether rat or human, it regurgitates fresh blood laden with plague bacilli into the bite wound, initiating the bubonic plague form of the disease. The bacilli can then reach densities of 1 billion individuals per milliliter of blood in the infected host.

The bacterium can also initiate pneumonic plague, spread directly between people through respiratory fluids, which nearly always results in death.[206] Rarely, the disease can spread from person to person via the bite of a human flea, which initiates septicemic plague, and this route is thought to lead to death within hours of infection.[206,207]

Yersinia pestis is perfectly adapted to initiate pandemics. It can persist for five weeks in the feces of rat fleas.[207] Some species of rat fleas can survive for over four months without food, and under ideal conditions may persist without hosts for close to two years.[272] *Yersinia pestis* is transmitted by more than eighty species of fleas and infects more than two hundred species of mammals, all of which can serve as a reservoir for disease outbreaks.[115,273] Analysis of ancient DNA from mass grave sites dated to the Justinian-era plague and the Black Death confirmed that both pandemics, like the third plague pandemic of the 1890s, were due to *Yersinia pestis*.[274–80]

The plague pandemic of the 1890s killed 15 million people during its rapid spread along modern transportation corridors. At the turn of the century, the pandemic spread from Asia throughout the world, hitting Africa, Europe, Australia, and North and South America—every occupied continent.[262] In some places people accused the scientists of initiating epidemics from their research laboratories, something that actually happened through negligence in Vienna.[281] Yersin faced such accusations in China, where an outbreak occurred near his lab in 1898.[262] The greatest devastation occurred in India, where 12.5 million people died of the disease between 1898 and 1918.[273] As had happened in Hong Kong, British attempts to control the epidemic in the Indian subcontinent were met by widespread rioting and violence.[262]

The advent of agriculture created a ready supply of food for rats and other rodents that carry the plague. Bubonic plague outbreaks occurred in the times of Justinian and Petrarch because rat fleas transmitted the disease from its natural reservoir—wild rodents, such as grass rats, gerbils, and marmots—to the domesticated black rat.[115] The infected flea moved along trade routes into Europe not only on the rat, but also on bundles of marmot fur.

After discovering how fleas transmit plague, Bacot played a key role in the 1914 Sierra Leone Yellow Fever Commission.[270] After a year in Sierra Leone, he focused his research efforts on antilice measures. These measures were key to keeping soldiers healthy in the trenches of World War I, where trench fever raged. He tested his antilice measures on himself in realistic field conditions and, along with his colleagues, demonstrated the role of lice in the transmission of trench fever. In 1920, Bacot inadvertently infected himself with trench fever, and used the opportunity to feed lice collected from a public bathhouse upon his blood. He faced a difficult fever, and the lice that imbibed his blood were infected with the microbe that he and his colleagues had earlier described. For months, he continued to infect experimental lice by feeding them upon himself. Then, in 1922, he joined an expedition to Cairo to study the role of lice in the transmission of typhus. There he added to the long list of other typhus investigators who died of the disease they were studying.

During Simond's 1898 experiments on rats, he deduced the important role of pesticides in combating plague. "The mechanism of the propagation of plague," he wrote, "includes the transporting of the microbe by rat and man, its transmission from rat to rat, from human to human, from rat to human and from human to rat by parasites. Prophylactic measures, therefore, ought to be directed against each of these three factors: rats, humans and parasites."[265]

Two of these three lines of attack could be achieved with pesticides—rodenticides against the rat and insecticides against the rat flea. To prevent the long-distance dispersal of plague, Simond recommended the fumigation of ships with sulfurous acid to kill the rat and its fleas.[247] This was a far cry from the prevailing and unsuccessful method of sanitation directed against humans as the mode of infection. And it eventually led to international conventions on the elimination of rats from ships and planes.[282]

Although Simond's experimental demonstration that rat fleas transmit plague was not internationally accepted until the 1906 English commission conducted its confirmatory experiments, Simond found more receptive ears among Indian authorities. Immediately after his rat experiments, in June 1898, the Indian government began "a vigorous poisoning crusade against rats."[260] Sanitary workers flooded sewers with phenol (carbolic acid), spread carbolic powder around houses, and pumped sulfur fumes into rat runs.[283] Rats were also poisoned with arsenic, phosphorus, barium carbonate, camphor, chloride of lime, extracts of the plant *Scilla maritima*, and strychnine.

Strychnine producers derived the poison from the seeds of trees in the genus *Strychnos*. Although strychnine had been used to poison animals for centuries, its identity as a chemical compound was not determined until 1818, when the French chemists Pierre-Joseph Pelletier and Joseph Bienaimé Caventou extracted it from a species of *Strychnos*.[284] Two years later Pelletier and Caventou completed their historic extraction of quinine and cinchonine from cinchona bark.[68]

In India, to disinfect just 180 houses, "the enormous quantity of 13,500 cubic yards of carbolic acid were used, quicklime was used in cartloads, and also lime with sublimate; liquid disinfectants were scattered about the houses by means of fire-bombs in such quantities that for some time afterwards, filtering through from one floor to another, they made necessary the use of umbrellas to those who repaired the houses."[281]

All these efforts met with poor success due to the inaccessibility of rats in sewer systems and underground nooks and crannies.[283] But they reduced public panic. "The streets were veritable pools of hypochlorite of lime and carbolic acid," wrote an observer, "the strong odour of which in some districts concealed that of misery and stinking fish, and raised at one and the same time the stomachs and the spirits of the public. Generally speaking, the authorities prefer disinfectants with a strong odour, because they inspire public confidence; the people who stop half suffocated at some corner by a smell like a druggist's shop believe in good faith that millions of bacilli which would otherwise have attacked them are dead at their feet, and they go back praising the zeal of the authorities."[281]

Other, more creative efforts were also employed in India, such as the inoculation of rats with a blood-poisoning pathogen followed by their release into rat-infested areas (even though the pathogen was also communicable to people), or the release of cats and screech owls in homes to hunt rats.[283] The problem with campaigns directed strictly against rats without simultaneous use of insecticides against fleas was that it encouraged fleas to leave their dead rat host in preference for a living human host.[282]

Not only scientists, but also common people, noted an association between resistance to plague and products that were also used as insecticides. An oil store manager in Bombay observed in 1903 that his workers who handled oil and were regularly covered with it did not contract plague, while several workers who did not come into contact with oil died of plague.[229] Similarly, the British consul in Egypt in 1797 reported that, although death from plague was widespread, no mortalities occurred among men working the oil fields. Similar anecdotes were common, such as this report of a British observer in India: "Another remark which the natives make, and which I think is likely to be just, as they are not apt to take notice of anything that is not extremely obvious, is, that those engaged in the expression of oil are not liable to infection."[229] Likewise, tobacconists who resided in their shops seemed to escape infection.

Bacot conducted an extensive experiment on the effectiveness of various insecticide vapors in killing rat fleas during the 1914 plague epidemic in India.[285] The arsenal of insecticides that Bacot tested included Lysol, flake naphthalene, formalin, benzene, paraffin oil, crushed camphor, ammonia, and phenol. Each of these compounds developed along its own path, which in the case of phenol became tragic.

Phenol was first extracted from coal tar in 1834 and called *karbolsäure*,[286] or coal-oil-acid, which also came to be known as carbolic acid. It developed into an important pesticide in the late nineteenth and early twentieth centuries. Tragically, its killing power was extended by the Nazis, who used phenol injections to kill thousands of concentration camp inmates during the Holocaust.[287]

This chemical crossover between pesticides and weaponry made its debut on a gigantic scale during World War I. The confluence of modern chemistry with industrial production technologies facilitated the arms race for superior chemical weapons and the transition from pesticides based on naturally occurring materials to synthesized compounds. It so happened that some newly synthesized pesticides made excellent chemical weapons, and innovations in chemical weapons design were readily translated into novel pesticides and delivery systems. The demands of war opened new frontiers in the fight against pests, and the status of chemists skyrocketed on both sides of the Western Front.

[**6**]

Synthetic Chemicals of War

(423 BCE–1920)

I feel it a duty to state openly and boldly, that if science were to be allowed her full swing, if society would really allow that "all is fair in war," war might be banished at once from the earth as a game which neither subject nor king dare play at. Globes that could distribute liquid fire could distribute also lethal agents, within the breath of which no man, however puissant, could stand and live.—**Dr. B. W. Richardson**, 1864[288]

The use of chemicals in combat did not start in World War I. More than two thousand years earlier, in ancient Greece, during the Peloponnesian War between Athens and Sparta in 423 BCE, the Athenians were unable to defend their fort at Delium against a fire and gas attack by the allies of Sparta. Fortunately for history, the events were recorded by Thucydides. Thucydides survived infection of plague during the Athens epidemic and fought as a general in charge of the naval forces of his region.[289] He described the attack on the fort at Delium, accomplished with a giant cauldron equipped with massive bellows to pump toxic flaming gas through an iron tube and wooden pipe: "The blast passing closely confined into the cauldron, which was filled with lighted coals, sulphur and pitch, made a great blaze, and set fire to the wall, which soon became untenable

for its defenders, who left it and fled; and in this way the fort was taken."[290] For his part in the separate loss of Amphipolis to the Spartans, the government exiled Thucydides for twenty years.[289] An assassin took his life around 400 BCE.

A thousand years later, "Greek fire" saw its debut in naval warfare. The probable constituents were plant resin, sulfur, naphtha (liquid hydrocarbon mixture), lime, and saltpeter (potassium nitrate).[288,291] The authoritative description of Greek fire, published in 1788, held that it "was most commonly blown through long tubes of copper, which were planted on the prow of a galley, and fancifully shaped into the mouths of savage monsters, that seemed to vomit a stream of liquid and consuming fire."[292] In the Middle Ages, a knight who faced Greek fire in the Crusades said that "it came flying through the air like a winged long-tailed dragon . . . with the report of thunder and the velocity of lightning; and the darkness of the night was dispelled by this deadly illumination."[292]

A more repulsive development came in 1680, when a chemist used extracts of human feces to try to fix mercury.[288] He accidentally produced a mixture called pyrophorus that, like Greek fire, burst into flame when exposed to the air. It was not until 1713 that another chemist discovered that human feces were not an essential ingredient of this modern version of Greek fire. Chemists refined the ingredients of pyrophorus such that a chemical reaction produced sulfurous acid that, when it encountered air, ignited into a great inferno.

Chemical warfare based on the scientific principles of the burgeoning field of chemistry, rather than based simply on trial and error, arose during the nineteenth century. During the Napoleonic Wars, the British naval officer Thomas Cochrane advocated for ships to be loaded with alternating layers of sulfur and charcoal on top of a bed of clay; the ships would be anchored near French battlements and lit when the winds were auspicious.[293] He argued that the ensuing sulfur dioxide gas would debilitate French forces and thereby facilitate a British incursion. The British Admiralty rejected Cochrane's proposal in 1812 due to uncertainties associated with wind, tides, and currents.

Cochrane persevered after becoming an admiral, and improved upon his plan by incorporating smoke screens in an 1846 proposal. "All fortifications," he wrote in his secret plan, "especially marine fortifications, can under cover of dense smoke be irresistibly subdued by fumes of sulphur kindled in masses to windward of their ramparts."[293] The plan was rejected again over perceptions that such an attack would not "accord with the feelings and principles of civilized warfare," and that it would invite counterattacks by the same method.

Cochrane resubmitted his proposal during the Crimean War in 1854.[294] He was by that time seventy-nine years old, and it was his last chance to initiate modern chemical warfare—a deployment he offered to personally oversee on the field of battle.[293] But the Ordnance Committee that reviewed the plan, including the scientist Michael Faraday, once more rejected the idea because smoke would likely not shield the ships and the enemy could counter with respirators. Faraday was eminently suited to review the science behind Cochrane's proposal; he had, after all, discovered charged atoms, which he called ions, and thereby initiated the new field of electrochemistry.[295] The Ordnance Committee concluded that Cochrane's plan was "hazardous, unpromising of success, and by probable failure likely to bring discredit on the service, and to give the enemy cause of boastful advantage calculated to improve his ebbing strength in the great struggle in which we are engaged."[293]

When the Aberdeen Coalition government fell from power due to poor handling of the Crimean War, Cochrane submitted his plan to the new government, which viewed it more favorably. Prime Minister Lord Palmerston agreed with Lord Panmure, the secretary of state for war, that the scheme should be implemented with Cochrane in charge. "If it succeeds," wrote Palmerston to Panmure, "it will, as you say, save a great number of English and French lives; if it fails in his hands we shall be exempt from blame, and if we come in for a small share of the ridicule, we can bear it, and the greater part will fall on him."[293] Before Cochrane could charge forth with his sulfur ships, however, Sebastopol fell and the war ended.

Cochrane's plan remained secret and was bequeathed to a family friend and then passed on to Cochrane's descendants. With the plan came explicit instructions that it could only be disclosed "in case of national emergency."[293] That emergency arrived in 1914. Cochrane's grandson brought the plan to the British Army and Navy. Although the army dismissed the plan, Winston Churchill at the Admiralty recognized its potential, but had no interest in violating the rules of war. Churchill did authorize experiments on the deployment of smoke screens to shield a chemical attack, and he placed Cochrane's grandson in charge. But the Germans struck with gas attacks first. The British then deployed Cochrane's smoke screens from ships near shore to shield the initiation of their own chemical attacks. Cochrane's grandson declared that chemical weapons were "the most powerful means of averting all future war" since they "would frighten every nation from running the risk of warfare at all."

Arguments for the humanity of chemical weapons had arisen during the Crimean War. Like Cochrane, a British chemist promoted cyanide use in artillery rounds against the Russians.[296] Officials at the War Office prohibited such weaponry on the grounds that it was akin to poisoning the enemy's water supply. "There is no sense in this objection," wrote the chemist. "It is considered a legitimate mode of warfare to fill shells with molten metal which scatters among the enemy, and produces the most frightful modes of death. Why a poisonous vapour which would kill men without suffering is to be considered illegitimate warfare is incomprehensible. War is destruction, and the more destructive it can be made with the least suffering the sooner will be ended that barbarous method of protecting national rights."[296]

Thus the contentious debate over chemical weapons began. Chemical weapons then attracted advocates and detractors on both sides during the American Civil War. US General Gilmore stimulated this debate when he launched shells with liquid fire upon Charleston. Confederate General Beauregard declared it to be "the most villainous compound ever used in war."[288] Advocates of chemical weapons thought otherwise. "I do not see that humanity should revolt,"

wrote one supporter of chemical weapons development during the Civil War, "for would it not be better to destroy a host in Regent's Park by making the men fall as in a mystical sleep, than to let down on them another host to break their bones, tear their limbs asunder, and gouge out their entrails with three-cornered pikes;—leaving a vast majority undead, and writhing for hours in torments of the damned? . . . War has, at this moment, reached, in its details, such an extravagance of horror and of cruelty, that it cannot be made worse by any art, and can only be made more merciful by being rendered more terribly energetic."[288] The US delegate to the 1899 Hague Conference to ban chemical weapons agreed, stating, "The reproach of cruelty and perfidy addressed against these supposed shells was equally uttered previously against fire-arms and torpedoes, although both are now employed without scruple. It is illogical and not demonstrably humane to be tender about asphyxiating men with gas, when all are prepared to admit that it is allowable to blow the bottom out of an ironclad at midnight, throwing four or five hundred men into the sea to be choked by the water, with scarcely the remotest chance to escape."[297]

Chlorine gas weapons were proposed but not used during the American Civil War. A New York schoolteacher suggested, in a detailed letter to the secretary of war, the use of chlorine gas delivered with projectiles against Confederate forces.[298] An equally creative northerner proposed the use of hydrogen chloride gas. "It has occurred to me," he wrote, "that Gen. Burnside, with his colored troops might, on a dark night, with a gentle breeze favorable, surprise and capture the strongholds of Petersburg, or Fort Darling, perhaps, without loss or shedding of blood."[299]

Such breakthroughs were suddenly possible because of rapid progress in organic chemistry that began with the accidental synthesis of urea from cyanic acid and ammonia in 1828 by the twenty-eight-year-old German chemist Friedrich Wöhler.[300–302] Before that momentous chemical reaction, scientists believed that matter was either organic or inorganic, and that organic matter could only be formed due to the vital force of living organisms.[295] Urea was a product

of the liver, and yet Wöhler synthesized it from inorganic matter in the lab. If that could be done, perhaps chemists could synthesize all organic materials. During the next few decades, not only did chemists synthesize a dizzying array of known organic matter, they also made many chemicals that do not occur in nature. Arising from these amazing achievements were new opportunities for chemical weaponry. These weapons sprang to life during World War I.

In 1914, Germany dominated the field of chemistry—in scholarly training, scientific achievement, range of chemical products, and industrial output.[303] Eight German chemical companies produced about 80 percent of the dyes for the world market. The intellectual drive behind Germany's supremacy in chemistry came in large part from the competitive efforts of two men, Fritz Haber and Walther Nernst.

The rivalry between these two great German chemists began despite, or perhaps because of, their similarities.[301] Nernst was only four years older than Haber. Both were short of stature. The two enjoyed the same cultural activities, and both served as investigating emissaries for Germany to the United States, Haber in 1902 and Nernst in 1904.

By that time Nernst was already wealthy thanks to his invention of an efficient light source that outperformed Thomas Edison's new carbon filament lamp.[295] Edison's light required a vacuum and produced weak lighting. Nernst's light employed a solid electrolyte containing cerium oxide and produced a bright light. The Kaiser was much impressed by Nernst's light, especially when Nernst showed him that, like a candle, he could light it with a match and extinguish it with a puff of air. This was because cool air made the cerium oxide nonconducting, while heat allowed it to pass an electric current.

After Nernst sold the patent for the astonishing sum of 1 million marks, he supervised an American graduate student named Irving Langmuir to study the behavior of an incandescent metal wire in a gas-filled glass bulb.[295] As a result, Nernst supervised the undoing of his own invention. Langmuir completed his studies with Nernst and took employment at the General Electric Company, where he

perfected the incandescent bulb into the replacement for the bulbs of both Edison and Nernst. Langmuir also discovered atomic hydrogen (isolated hydrogen atoms) and won the 1932 Nobel Prize in Chemistry.[295,304]

Nernst invested his lightbulb money in his laboratory, in addition to acquiring personal luxuries.[295] In 1898, he bought the first car in Göttingen, which was the first of eighteen that he eventually owned. One car even had its boiler under the driver's seat and issued trails of flame out the exhaust. Nernst could not resist the temptation to investigate the oxidation of petrol in the internal combustion engine and to improve it. He calculated that the heat output could be augmented by injecting nitrous oxide into the cylinder, and so he had a nitrous oxide gas tank added to his car. He then employed the gas to ascend a hill too steep for automobiles, all the while pressing on his horn as pedestrians scattered.

In short order, Nernst obtained a professorship at the University of Berlin.[295] As an innovator at the cutting edge of technology, he decided to drive there with his family in his personal car. The car broke down en route. Although Nernst developed the thermodynamic theory for the nature of galvanic elements, apparently he had charged his car's battery using the wrong polarity.

Nernst's colleagues described how one day God set out to create a superman. "He began his work on the brain and formed the most perfect and subtle mind, but then he was unfortunately called away. The archangel Gabriel saw the unique brain and could not resist the temptation to shape the body, but unfortunately, due to his inexperience, he only succeeded in fashioning a rather unimpressive looking little man. Dissatisfied with his efforts, he left his work. Finally, the devil came along and saw the inanimate thing, and he blew the breath of life into it."[295]

The facilitation of Nernst's productivity came not from the devil, but from his wife Emma. Emma ran the logistics of their household with five children to maintain their close-knit family life.[295] She entertained for the active social life with scientists and luminaries that revolved around her husband. She typed while Nernst dictated

his manuscripts. She even inspected the starched cuffs of his shirts before they were laundered because he often jotted notes on them that had to be transcribed before they were wiped clean by soap and water.

Nernst was the most famous physical chemist in Germany.[301] When Haber sought a professorship at Leipzig, Nernst successfully blocked the appointment. Some of Haber's frustrated attempts at promotion (though not thwarted at the hand of Nernst) were a by-product of antisemitism; to facilitate his career trajectory, Haber converted to Protestant Christianity. A colleague later stated that "before thirty-five he was too young for a professorship, after forty-five he was too old, and in between he was a Jew."[295]

Some of Haber's roadblocks were due to the inevitable setbacks of a rapidly developing field where competitors scrambled to scoop each other. Haber wrote in his book *Thermodynamics* that a temperature of zero degrees Kelvin could not be achieved, but it was Nernst who had the insight (midway through a lecture) that allowed him to develop this into the third law of thermodynamics.[295] Nernst thus gained the stature, attained by only a select few, of developing a scientific law. Nernst said, "The first law of thermodynamics rests on the shoulders of many; the second law on the shoulders of few; the third, on the shoulders of one—mine."[301] Scientists complained that to find text about the new law in Nernst's book, they had to search for the entry "my heat theorem" in the book's index.[295]

Haber's creativity also seemed boundless, and it was tethered to a sense of humor. In his bachelor days, he and his close friends, among them several artists, gathered at their regular table, over which was affixed a horn and shield inscribed with the phrase, "At this table a few lies are permitted."[305] Haber told his friends the tale of the village well in Alvaneu, where, after a long walk on a hot day, he arrived thirsty and at the same time as a huge ox. He and the ox both plunged their heads into the cold water of the well and thereby got their heads mixed up.

Haber's productivity, energy, and attraction to insurmountable problems also attracted collaborators from all over the world to his

lab, sometimes more than forty at a time. Haber wrote to his friend, the Nobel Laureate Richard Willstätter, "I am delighted with everything that exceeds my ability and am happy when I can admire."[305]

Haber and Nernst both sought to achieve the seemingly unachievable: the fixation of atmospheric nitrogen into ammonia usable for fertilizer.[301] Such an accomplishment would do a great deal to alleviate world hunger, since agricultural production was limited by the availability of mined nitrate deposits from Chile, which at the turn of the century supplied two-thirds of the world's nitrogen for fertilizers. Haber's initial experiments yielded only minute quantities of ammonia, despite the use of conditions such as a reaction temperature of 1,000°C. Nernst concluded that Haber had miscalculated the yield of ammonia in his experiments. Nernst revised the equations and set out to be the first to exact an economically achievable yield of ammonia from atmospheric nitrogen. This intensified the bitter rivalry between Haber and Nernst, each aspiring to be the greatest physical chemist in Germany, and hence the world.

Both Nernst and Haber realized that increasing pressure during the chemical reaction could improve the yield of ammonia.[301] This logic was based on Henry Le Châtelier's law on chemical equilibrium. In 1901, Le Châtelier used elevated pressure to synthesize ammonia, which led to a terrible explosion that nearly killed his assistant, but also improved the understanding of the role of pressure in chemical reactions. Both Nernst and Haber experimented with high pressures, in the case of Nernst's studies up to seventy-five times higher than atmospheric pressure. Nernst was the first to successfully synthesize ammonia with the aid of high pressure, but the quantities were small. Haber consistently calculated a yield of ammonia about 50 percent higher than Nernst's calculations for the same reaction. Their intellectual rivalry came to a head at the 1907 meeting in Hamburg of the Bunsen Society, where Haber's reputation was on the line for challenging Nernst and his scientific standing. But who was right and who was wrong in his calculations was still unknown.

Two years later Haber and his students used 200 atmospheres of pressure and 600°C to achieve an 8 percent yield of ammonia.[301] To do this, they had to construct a reaction chamber capable of enduring such extreme pressure and temperature, and they had to find a catalyst to facilitate the reaction. Haber discovered that the elements uranium and osmium successfully catalyzed the reaction and that a metal chamber could withstand the necessary conditions. He then negotiated an unusual deal with an industrial partner whereby he would receive one pfennig for every kilogram of ammonia sold so that price fluctuations or improved manufacturing efficiency would not have an impact on his income. Haber had solved the world's fertilizer problem and thereby initiated the Green Revolution. Nernst continued to lecture that Haber's innovation in the synthesis of ammonia really belonged to him, Nernst. The feud led Haber and Nernst to refuse to attend conferences if the other was invited.

Haber's fixation of nitrogen put him at the forefront of German chemistry.[301] Haber was, as many historians of science would argue, the most influential chemist in history.[306] Without nitrogen fixation, perhaps one-third of the world's rapidly expanding population in the twentieth century would have starved.[307] Many of the best young chemists in the world sought out training with Haber rather than with Nernst, and Haber's lab produced some of the most famous scientists of the time.[301] Haber was provided with his own research institute, the Kaiser Wilhelm Institute for Physical Chemistry and Electrochemistry.

Haber's innovative methods seemed without limit. In response to a request from the Kaiser for an effective indicator of the poisonous firedamp gas encountered in coal mines, Haber invented a whistle that changed sound when too much methane was present in the mine's air and that produced a distinctive trill when the gas level was sufficiently high to explode.[301]

Haber's Jewish heritage no longer seemed to impede his career. Indeed, despite prominent antisemites in German society, the Kaiser did not share this prejudice.[295] By the time World War I broke out, three of the Kaiser's science institutes were directed by Jews

who would later win the Nobel Prize: Fritz Haber (1918 Nobel Prize in Chemistry), Richard Willstätter (1915 Nobel Prize in Chemistry), and Albert Einstein (1921 Nobel Prize in Physics).[295]

The Kaiser's popularity surged with the start of the war. Before an adulatory crowd he pronounced, "I shall lead you forward to glorious times."[295] For Nernst, the first manifestation of that glory was the loss of one of his two sons in combat.[295] In grief, Nernst joined the voluntary drivers' corps and delivered documents from the General Staff in Berlin to von Kluck's Second Army in France. Within two weeks, Nernst with von Kluck's forces reached the outskirts of Paris. But a sudden retreat followed their sudden advance, Nernst and his car were nearly captured by the pursuing French, and then the war bogged down in the trenches surrounding No Man's Land.

By Christmas, just five months into the war, Nernst told his family and friends that the war was lost.[295] This was an analytical judgment by a scientist who studied the dynamics of reactions. Germany's generals had planned on quick victory. They had no plan for protracted defense against enemies on all sides that were growing in strength while Germany consumed its limited resources for war. Reluctantly, the generals concluded that they had to turn to scientists to stave off defeat.

The German government assigned Nernst the task of devising effective chemical agents to use in weapons.[301] According to an American history of gas warfare published three years after the war, the idea to employ gas weapons was Nernst's.[297] Nernst suggested a combination of an irritant called dianisidine chlorosulphonate and a tear maker called xylyl bromide, but the chemical mixture did not produce a strong enough effect.[301] Historical records do not reveal whether the formula for a more lethal chemical weapon escaped the great mind of Nernst, or whether he was unwilling to produce one.[295]

Meanwhile, Haber solved his assignment to produce an effective gasoline antifreeze to enable the German military to fight through the Russian winter. Given this success, the government replaced Nernst with Haber to lead the poison gas research. The Kaiser appointed Haber as the director of the Headquarters for Research and

Testing in Gas Warfare and Poison-Gas Protection at the Ministry of War.[308]

Nernst was reassigned to the task of developing new explosives.[295] For one of his tests, he decided not to bother driving to the proving grounds, but instead placed the explosive in the bottom of an unused well next to his campus laboratory. He reasoned that the blast would proceed upward and hence would pose no risk. Unfortunately, the well opened at the bottom into ventilation shafts that provided air for adjacent lecture halls. The explosion and subsequent darkness in the main lecture room, now filled with clouds of dust, startled the chair of the physics department and his three hundred students.

For his new assignment, Haber led the poison gas research at his institute.[301] His team chose chlorine because large volumes of locally available liquid chlorine could be stored under pressure in cylinders, and when it was released, the gas, being heavier than air, would hover near the ground.[295] The team experienced a setback in December 1914 when an explosion killed one leading researcher (a close friend of Haber's wife Clara) and injured another.[301,309] By the following month, however, Haber completed the critical research on the best humidity and wind conditions for deployment. He discovered that if he could see the movement of grass, then the wind was too strong for an attack.[295] Only a gentle breeze would suffice.

Haber was convinced that chlorine gas could be an effective weapon.[301] Indeed, in a test release, Haber himself suffered a grave injury from accidental chlorine gas exposure.[310] Germany had signed the 1899 Hague Convention banning the use of asphyxiating gases, but in the midst of a military stalemate the Germans chose to ignore their previous intentions, which suddenly seemed naïve.[301] Russia had already tried to use chlorine gas, but the cold weather at the time had driven the chlorine down into the snow. The chlorine revolatized the following spring, by which time German forces were far away.

On April 22, 1915, Haber supervised the release of 150 tons of chlorine gas from 5,730 gas cylinders near Ypres in Belgium in the first use in history of a weapon of mass destruction.[301,310] A breeze conveyed the gas toward enemy trenches. French Algerian soldiers

retreated, but French territorial and Canadian troops surged forward into the gas. An observer wrote: "Try to imagine the feelings and the condition of the colored troops as they saw the vast cloud of greenish-yellow gas spring out of the ground and slowly move down wind towards them, the vapor clinging to the earth, seeking out every hole and hollow and filling the trenches and shell holes as it came. First wonder, then fear; then, as the first fringes of the cloud enveloped them and left them choking and agonized in the fight for breath—panic. Those who could move broke and ran, trying, generally in vain, to outstrip the cloud which followed inexorably after them."[297]

A German soldier who helped to release the gas wrote:

> We should have been going to a picnic, not doing what we were about to do. The artillery put up a really heavy attack, starting in the afternoon. The French had to be kept in their trenches. After the artillery was finished, we sent the infantry back and opened the valves with strings. About supper time, the gas started toward the French, everything was quiet. We all wondered what was going to happen. As this great cloud of green gray gas was forming in front of us, we suddenly heard the French yelling. In less than a minute, they started with the most rifle and machine gun fire that I had ever heard. Every field artillery gun, every machine gun, every rifle that the French had must have been firing. I had never heard such a noise. The hail of bullets going over our heads was unbelievable, but it was not stopping the gas. The wind kept moving the gas towards the French lines. We heard the cows bawling, and the horses screaming. The French kept on shooting. They couldn't possibly have seen what they were shooting at. In about fifteen minutes, the gun fire started to quit. After a half hour, only occasional shots [were heard]. Then everything was quiet again. In a while it had cleared and we walked past the empty gas bottles. What we saw was total death. Nothing was alive. All of the animals had come out of their holes to die. Dead rabbits, moles, rats, and mice were everywhere. The

smell of the gas was still in the air. It hung on the few bushes that were left. When we got to the French lines, the trenches were empty. But in a half mile, the bodies of French soldiers were everywhere. It was unbelievable. Then we saw that there was some English. You could see where men had clawed at their faces, and throats, trying to get their breath. Some had shot themselves. The horses, still in the stables, cows, chickens, everything, all were dead. Everything, even the insects were dead . . . All of us went back to our camps and quarters wondering what we had done. What was next? We knew what happened that day had to change things.[311]

A clergyman, watching horror-stricken through his field glasses, described what he saw happen to Allied soldiers: "a greenish-gray cloud had swept down upon them, turning yellow as it traveled over the country, blasting everything it touched, shriveling up the vegetation. No human courage could face such a peril. Then there staggered into our midst French soldiers, blinded, coughing, chests heaving, faces an ugly purple color—lips speechless with agony, and behind them, in the gas-choked trenches, we learned that they had left hundreds of dead and dying comrades."[297] As many as ten thousand men were injured and between five and ten thousand died.[301,311]

The Germans did not take advantage of the break in the Allied line to launch a large-scale attack.[310] They were unprepared to do so because commanders on the front line could not overcome their suspicion of a civilian scientist telling them what to do in battle.[295] Nor could they give much credence to a weapon that required particular conditions of weather. This frustrated Haber, who said, "Early in 1915, gas was employed by the German and French sides in small quantities with no results. Then, while we were experimenting with a liquid commonly called a gas because it became effective when it vaporized, I advocated massed gas attacks to break the war's stalemate. But I was a college professor, and therefore not to be heeded by the leaders. They admitted afterward that if they

had followed my advice and made a large-scale attack instead of the experiment at Ypres, the Germans would have won."[301] The chief of the Chemical Warfare Service in the United States agreed with this assessment.[297]

The parallels between gas warfare and pesticide applications were not lost on the combatants. At Ypres, a German general said, "I must confess that the commission for poisoning the enemy just as one poisons rats struck me as it must any straightforward soldier; it was repulsive to me."[303] Within a day of the attack at Ypres, the British commander Sir John French telegrammed to London: "Urge that immediate steps be taken to supply similar means of most effective kind for use by our troops."[312] Thus began the tit-for-tat gas attacks that characterized World War I.

Haber's wife, Clara, herself a talented chemist before giving up her career to marry, viewed poison gases as barbaric.[295,301,309] She argued, pleaded, and demanded that her husband disown them. Although Haber felt traumatized after supervising the first gas attack at Ypres, he believed that poison gases could ensure a speedy victory for Germany. He explained to Clara that a scientist works for the world during peaceful times but for his country during war. On May 1, 1915, Haber and his colleagues celebrated the successful Ypres attack in his institute director's mansion.[309] That night, Clara went outside into the garden and shot herself with Haber's service revolver. Their thirteen-year-old son Hermann found her as she lay dying. Her suicide was apparently motivated by a multitude of factors in addition to her husband's development of chemical weapons: her unfulfilled career as a chemist, the death of close friends, and Haber's dalliance with his future wife. Later that day, May 2, Haber returned to duty on the Eastern Front.

After fighting on both the Western and Eastern Fronts, and getting wounded on the battlefield, the future Nobel Prize winner James Franck was assigned to Haber's institute in Berlin.[308,314] Franck, along with Gustav Hertz, provided the first experimental support of Bohr's theory of atomic structure; when he lectured on the discovery, Einstein said, "It's so lovely, it makes you cry!"[314] At

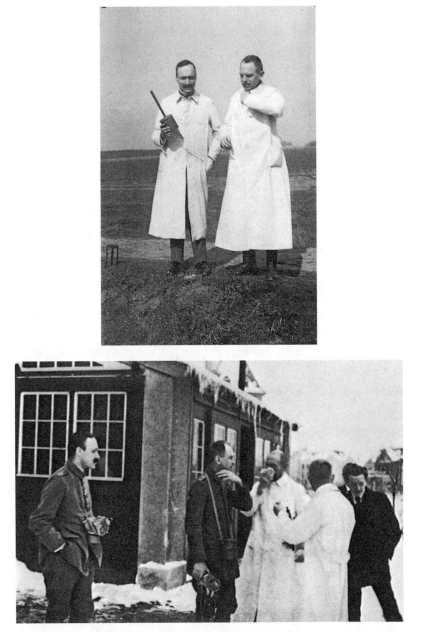

Figs. 6.1a and 6.1b. James Franck (left in both photos) and Otto Hahn (on Franck's left in both photos) testing German gas weapons and masks (visible in bottom photo) during World War I. The bottom photo shows the house in which they tested the efficacy of gas masks at Fritz Haber's institute in Berlin.[313]

Haber's institute, Franck tested the efficacy of gas masks and filters, and he served as Haber's confidential assistant at the front. Other scientists in the testing crew included Otto Hahn (1944 Nobel Prize in Chemistry for the discovery of nuclear fission), Gustav Hertz (1925 Nobel Prize in Physics with Franck, and the nephew of Heinrich Hertz, who demonstrated the existence of electromagnetic waves),[313] and Hans Geiger (invented the Geiger counter and, under Ernest Rutherford's direction, experimentally demonstrated that atoms have a nucleus). Filter design was the responsibility of the Nobel Laureate Richard Willstätter, director of the Kaiser Wilhelm Institute for Chemistry.[310] The scientists donned masks within sealed rooms full of the poison gas and stayed there until they noted that the mask and filter were no longer effective. Given that they did not know what length of exposure was fatal, the research was highly risky.

That Haber's subordinates served as guinea pigs for tests of the efficacy of new gas mask technologies followed an age-old tradition. Gas masks were invented at least as early as 1854. "The important agent in this instrument is the charcoal," read a description of the mid-nineteenth-century respirator, "which has so remarkable a power of absorbing and destroying irritating and otherwise irrespirable and poisonous gases or vapors that, armed with the respirator, spirits of hartshorn, sulphuretted hydrogen, hyrdosulphuret of ammonia and chlorine may be breathed through it with impunity, though but slightly diluted with air. This result, first obtained by Dr. Stenhouse, has been verified by those who have repeated the trial, among others by Dr. Wilson, who has tried the vapors named above on himself and four of his pupils, who have breathed them with impunity."[297] During World War I, even horses, dogs, and carrier pigeons used on the front lines wore gas masks.[315]

On July 12, 1917, with chlorine gas attacks stymied by widespread use of improved gas masks containing neutralizing chemicals, the Germans launched artillery shells of dichlorethyl sulfide (mustard gas) at British forces at Ypres.[291,297] Mustard gas, developed at Haber's institute, was persistent, making both exposed air

and objects into weapons. It blistered the skin, with the burns emerging four to twelve hours after exposure.[297] During the first six weeks of the German bombardment with mustard gas, British forces experienced nearly twenty thousand casualties. Clearly, the Allies had to develop their own mustard gas.

The Allies' mustard gas was not battle ready until a year after the first German use.[301] The Allied success at synthesizing mustard gas was credited to the English chemist William J. Pope and to dye manufacturers in France and England. Mustard gas became the most important poison of the war. In one night of battle at Nieuport, the combatants launched more than 50,000 artillery shells, each containing up to 3 gallons of mustard gas.[297] Scientists then developed numerous new chemical agents and toxic combinations that were weaponized, including phosgene, also created at Haber's institute.[295] By war's end, one-quarter of artillery shells contained chemicals; the chemical weapons plants were churning out their maximum production capability.[316] Haber wrote to a colleague that conventional war fought with artillery was akin to a game of checkers, while gas warfare was a game of chess.[312]

War Gases and Insecticides (1914–20)

The disapproval which the knight had for the man with firearms is repeated by the soldier armed with steel weapons, against the man who opposes him with chemical weapons. The aversion, which arises from unfamiliarity with the weapon, is further increased by the appearance of exceptional ruthlessness and by the feeling that it may violate the fundamentals of international law, which must even in war remain sacred in the interest of civilization. After the ravings of the foreign press, which during the war could not judge the subject impartially but only from national prejudice, the truthful verdict can only be brought to light slowly.
—**Fritz Haber**, 1920[316]

Near the end of the war, for his achievements in the development of the trench mortar and the tests he conducted on both the Western

and Eastern Fronts, Nernst received the Kaiser's highest decoration in a German order that had a fixed limit on membership; in this distinction, he succeeded the just-deceased Count Zeppelin.[295] Around the same time, Nernst's remaining son died in battle. Nernst sought solace in his research, and opened his famous publication on thermodynamics with the sentence, "Nothing is as good as physics to divert the mind from the present time which, in spite of the greatness achieved by our people, is nevertheless to be deplored."[295]

Earlier, Nernst had tried to stop the madness. Like Haber, Nernst had a personal friendship with the Kaiser.[295] Nernst used that friendship to obtain an audience with the Kaiser and the two leaders of the German war effort, Paul von Hindenburg and Erich Ludendorff. Nernst argued that unrestricted submarine warfare would draw the United States into the war and that this would create a resource imbalance that Germany could not overcome. Ludendorff interrupted Nernst and dismissed his analysis as inept gibberish by a civilian.

In 1914, American chemical companies produced only simple organic chemistry products and relied upon chemical reactants purchased from Germany.[303] Germany's chemical production was twenty-one times greater than that of the United States.[317] The war would change all that, leaving behind a battered and impoverished Germany and an emergent and powerful chemical industry in the United States.

In 1917, seventeen American companies began to produce dyes to fill the void left by the blockade and collapse of German industry.[303] Two of these companies, DuPont and National Aniline & Chemical (subsequently Allied Chemical), surged to the forefront of the chemical marketplace. Another American firm, Hooker Company, produced only bleach and caustic soda in 1914. By the war's end, it manufactured seventeen chemicals, and dominated the world production of monochlorobenzol, used to produce dyes, explosives, and poison gases. From 1914 to 1919, the annual value of American chemical manufactured products increased from $200 million to $700 million.[317] Just a few months after the war's

end, the United States produced enough poison gases to have dispatched 200 tons per day into the battlefield.[303]

This was a far cry from the situation in 1915, when a daring American idea was seriously proposed that, in retrospect, would seem farcical. "One ingenious person suggested" to the US Board of Ordnance and Fortifications "a bomb laden to its full capacity with snuff, which should be so evenly and thoroughly distributed that the enemy would be convulsed with sneezing, and in this period of paroxysm it would be possible to creep up on him and capture him in the throes of the convulsion."[297]

One-third of the seventeen thousand chemists in the United States worked for the federal government on the war effort.[318] The research arm of this expertise included seventeen hundred scientists, which was the greatest research group that the United States had ever assembled. The secretary of war said that no other profession was "more essential to our national success than that of the chemists . . . the chemical mind was at the highest tension and was emitting sparks of great luminosity to the very end."[318] Midway through the war, the conflict was aptly described as "a struggle between the industrial chemical and chemical engineering genius of the Central Powers and that of the rest of the world. Quite irrespective of the war's origin, aims, ideals or political circumstances, these are the cohorts from which each side derives its power."[303]

The war accelerated the organization of science in the United States in the civilian sector as well. The 1916 torpedo attack by a German U-boat on a French ship with American passengers motivated the US National Academy of Sciences to put its resources toward the war effort should America join the war.[303] The National Academy then set up the National Research Council, which has been a mainstay of scientific research for federal priorities ever since.

Not only did the war motivate chemists to develop poison gases, it also drove chemists to develop insecticides.[303] Demand for cotton far outstripped supplies. Cotton was needed for uniforms, tents, bandages, and propellants for explosives. Powder and explosives were derived from nitrocellulose (composed of cotton and nitric

acid), nitroglycerin (composed of animal and plant fats and nitric acid), and nitrotoluol (derived from coal tar and nitric acid).[316] Haber's fixation of atmospheric nitrogen to ammonia solved the nitric acid problem for Germany, which prevented Germany's collapse after natural stores of ammonia derived from Chilean saltpeter were exhausted early in the war. Synthetic ammonia production via the Haber-Bosch process grew from 6,500 tons of bound nitrogen per year in 1913 to 200,000 tons per year during the war.[305] But the availability of synthetic ammonia did not solve the problem of scarcity of fats and cotton. Military uses robbed the public of civilian uses. "When limited to our own sources of fat our nutrition suffered," wrote Haber, "even though no food fats were withdrawn for chemical uses. When fat was used for the preparation of glycerin we starved twice as much."[316]

Cotton was in short supply for both sides in the war. The boll weevil infestation overwhelmed US cotton farmers at the same time that the war vastly increased demand.[303] The damage done to cotton by boll weevils, especially in Georgia and South Carolina, became a major cause of the Great Migration of black Americans from the agricultural southern states to the industrialized North.[319] After successful testing, researchers settled on calcium arsenate as the solution to the boll weevil problem, and the chemical found widespread use beginning in 1917.[303] By 1920, twenty companies in the United States produced 10 million pounds of calcium arsenate per year.[320]

Calcium arsenate was based on arsenic, long known to be poisonous. Other arsenic-based pesticides, such as lead arsenate and Paris green, were already used extensively in the cultivation of fruits, timber, and potatoes.[303] Pyrethrum, derived from chrysanthemum flowers, was also widely used to kill insect pests. But no synthetic organic pesticides had yet been developed.

Chemical research on effective explosives and war gases led to the first synthetic organic pesticide. Chemical companies produced large quantities of picric acid for explosives.[303] A byproduct of the process was paradichlorobenzene, also called PDB. Entomologists tested PDB and various war gases for their toxicity to insects, with

promising results. PDB became the first of these synthesized war chemicals to enter the marketplace as an insecticide. PDB production for pesticides grew to millions of pounds per year by the early 1940s.

Chemists and entomologists also tested war gases for their effectiveness against body lice; because they transmitted typhus, body lice were a priority target in the war effort. The aim was to find "a gas which can be placed in a chamber and be experienced safely for a short period of time by men wearing gas masks and which in this time will kill all cooties and their nits."[303] The US Chemical Warfare Service, the Bureau of Entomology, and other collaborating government agencies tested the effectiveness of a suite of war gases against lice and other insect pests.[303] The most commonly used gas in the war, chloropicrin (trichloronitromethane, also called "vomiting gas") was popular among military strategists because it infiltrated gas masks and caused soldiers to vomit and tear up. Soldiers tore off their gas masks in response, and then inhaled lethal gases that had been mixed with chloropicrin. Chloropicrin also turned out to be an effective insecticide.

Research and development also proceeded the other way around, from insecticides to war gases. Allied chemists, particularly the French, intensively studied the utility of hydrocyanic acid (also called hydrogen cyanide or prussic acid) as a war gas.[297] The idea for use of hydrocyanic acid came from insect control, since it had been used to fumigate trees in orchards, which were first covered in tents, as well as buildings since the nineteenth century.[303,321] The gas had a low density, so researchers mixed it with other chemicals to keep it hovering near the ground.[297] The hydrocyanic acid mixtures were called vincennite and variously included chloroform, arsenic trichloride, and stannic chloride. Vincennite mixtures were heavily used in gas bombardments by the French, but were eventually abandoned in favor of other gas weapons.

The integration of arsenic into gases, based on the insecticidal power of arsenical compounds, became widespread.[303] This use consumed over one-third of the available arsenic in the United States, and thereby hampered the pesticide industry.[322] The most powerful

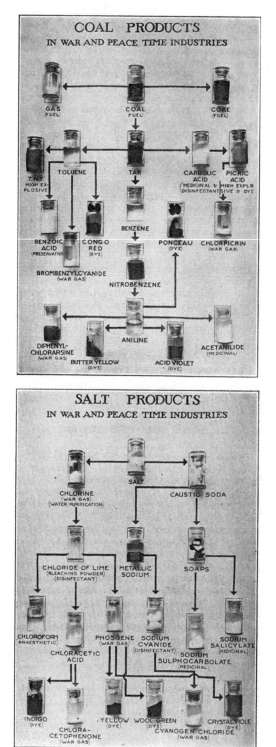

Figs. 6.2a and 6.2b.
The synthesis of useful products from coal (top), including the war gas and insecticide chloropicrin; the synthesis of useful products from salt (bottom), including chloroform used in vincennite and the war gas and rodenticide phosgene.[297]

of these arsenical compounds was lewisite, named after its American discoverer, W. Lee Lewis.[297] It arrived too late in the war for decisive use on the battlefield, but its combined properties as a blistering agent, respiratory irritant, and sneeze inducer, along with its lethality, led the chief of the US Chemical Warfare Service to label it the "dew of death." Three drops were enough to kill a rat.

The secrets of the synthesis and effects of lewisite were jealously guarded, but speculation abounded. The *New York Times* broke the story in 1919 and reported that workers at the lewisite manufacturing plant near Cleveland, dubbed the "Mouse Trap," could not leave the 11 acre compound until the war was won.[323] The paper reported that "ten airplanes carrying 'Lewisite' would have wiped out, it is said, every vestige of life—animal and vegetable—in Berlin . . . What was coming to Germany may be imagined by the fact that when the armistice was signed 'Lewisite' was being manufactured at the rate of ten tons a day. Three thousand tons of this most terrible instrument ever conceived for killing would have been ready for business on the American front in France on March 1."[323]

During the war, the battle lines against Germans and insects were purposely blurred.[303] America's leading entomologist, Stephen A. Forbes, wrote in 1917 that the United States had been invaded by "fifty billion German allies"; the "cinch bug is pro-German in our present war, the Hessian fly is still Hessian, and the army worm is an ally of the German army."[303]

The fighting in World War I resulted in more deaths than in all the nineteenth-century wars combined.[303] Approximately half of the deaths occurred among servicemen (about 10 million soldiers).[325] Of these deaths, about 90,000 were caused by chemical weapons, and another 1.3 million soldiers suffered gas injuries.[303] Although chemical weapons proved far less fatal than conventional weapons in the grand scheme of things, they instilled a visceral reaction that somehow was not triggered by bayonets, cannons, and machine guns—which, after all, were weapons familiar to everyone. Yet the death rate from gas attacks was relatively modest compared

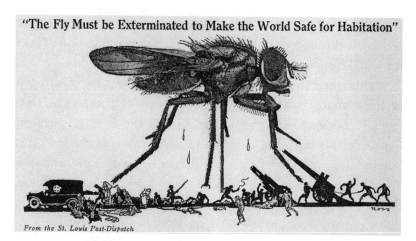

Fig. 6.3. Newspaper cartoon from the *St. Louis Post-Dispatch*, July 1918, with soldiers fighting a fly, a reference to the German enemy.[324] Casualties mount in the scene while refugees flee.

to that of these conventional weapons. For example, fewer than 2 percent of American gas casualties died, while more than 25 percent of casualties from bullets and bombs died.[297]

A leading American proponent of chemical weapons wrote that gas was "at one and the same time the most powerful and the most humane method of warfare ever invented."[303] The chemist who led the American mustard gas research and developed the new poison gas lewisite said, "To me, the development of new and more effective gases seemed no more immoral than the manufacture of explosives and guns . . . I did not see . . . why tearing a man's guts out by a high-explosive shell is to be preferred to maiming him by attacking his lungs or skin."[303] Haber agreed with this assessment.[316]

These statements on the part of the chemists who developed the weapons did not fall into line with public opinion. A typical newspaper article asked, "What is this 'humane' method of warfare of which the chemists speak? Is it the spreading of gas that will torture and poison honorable and gallant men not only through their lungs but through their skins, that will reach far behind the fighting lines and send women and children to horrible death, that will kill all

vegetation and secure the starvation of peoples for years after war ceases? If this be a chemist's idea of humane warfare, God deliver the world from its chemists!"[303]

By the final year of the war, the Haber process to fix nitrogen produced more than 200,000 tons of nitrogen compounds for Germany.[301] Midway through the war, one of the German nitrogen fixation plants stretched alongside nearly 2 miles of railway.[295] This industrial output played a critical role in Germany's perseverance despite the entrance of the United States into the theater of war. Nitrogen was a key component in all explosives, so it seems likely that Germany would have folded much sooner without Haber's innovation. In fact, Haber calculated that Germany would have lasted only until the spring of 1915 without it.[316]

Immediately after the war, both Haber and Nernst received Nobel Prizes in Chemistry for their fundamental discoveries—in 1919 Haber was awarded the 1918 prize for fixing atmospheric nitrogen into ammonia, and Nernst received the 1920 prize for discovering the third law of thermodynamics.[301] But Haber's global reputation was scarred by his association with gas warfare. So strong was this contempt that French scientists refused to accept the Nobel Prize because theirs was to be awarded at the same ceremony as Haber's. The *New York Times* supported the French scientists and their horror at the prize going to Haber. "One may wonder, indeed," wrote the paper, "why the Nobel prize for idealistic and imaginative literature was not given to the man who wrote General Ludendorff's daily communiqués."[326] A French scholar protested that Haber was "morally unfit for the honor and material benefits of a Nobel prize."[301]

Haber had the unhappy distinction of receiving the only Nobel Prize in the sciences that was ever contested. The global scientific collaborations that Haber had so diligently promoted before the war were now thwarted by his infamy. Where the brightest minds had competed for a spot in his institute before the war, they now refused to shake his hand in scientific gatherings. Haber felt, above all, that he must once again achieve the scientifically impossible to regain his global stature.

[**7**]

Zyklon

(1917–1947)

All modern weapons, although they appear to have the purpose to kill the enemy, owe their success finally to the vigor with which they overpower the morale of the enemy. The battles which decide the outcomes of the wars are not won through the physical destruction of the enemy, but through psychical concussions which overcome his powers of resistance in a deciding moment and bring up the picture of defeat. The troops, which are a sword in the hands of a leader, are made into a mob of desperate human beings by those psychical concussions.
—**Fritz Haber**, 1920[316]

As for Poison Gas and Chemical Warfare in all its forms, only the first chapter has been written of a terrible book.—**Winston Churchill**, 1932[327]

After World War I, the use of chemical weapons shifted to colonial subjugation of indigenous populations.[291] In 1920, the British deployed mustard gas against Afghans, in 1925 the Spanish did the same to Moroccans, and in 1935 Mussolini's forces repeated the tactic against Ethiopians. In the case of Ethiopia, Emperor Haile Selassie's forces were routed, and he complained to the League of Nations, "The deadly rain that fell from the aircraft made all those

whom it touched fly shrieking with pain . . . In tens of thousands the victims of Italian mustard gas fell."[315]

Nevertheless, most of the chemical weapons expertise and infrastructure built during World War I was directed to peacetime profitability through the production of both chemical weapons and pesticides. World War I necessitated the blurring of the lines between military and industrial activities. In the United States, the army's Chemical Warfare Service brought together the nation's best chemists to develop its chemical weapons arsenal.[303]

The first director of the Chemical Warfare Service, William Sibert, recognized that chemical weapons were here to stay. "History proves," he wrote three years after the war, "that an effective implement of war has never been discarded until it becomes obsolete."[297] Sibert was concerned that the expertise assembled by the Chemical Warfare Service would not be cultivated. "I feel that . . . the genius and patriotism displayed by the chemists and chemical engineers of the country were not surpassed in any other branch of war work and that to fail to utilize in peace times this talent would be a crime."[297]

Experts at the Chemical Warfare Service reasoned that chemical weapons, which were so effective at debilitating and killing people, would perhaps be even better at killing insects.[303] In the postwar period, chemists continued to test war gases against insect pests. Part of the motivation for this was political. The Chemical Warfare Service was slated for dissolution once hostilities ceased, as specified in President Wilson's order that created it.[315] Having survived this existential threat, the Chemical Warfare Service faced budget cuts from Congress as the need for chemical weapons seemed to be a thing of the past.[303]

To maintain its relevancy, the Chemical Warfare Service gave itself a makeover and promoted the civilian benefits of its poison gases, especially their potential uses as insecticides. It began to call itself the Chemical Peace Service, conducting "peaceful warfare."[303] "It is very possible," noted one analysis, "that our investigations will demonstrate that the quickest and surest method in attacking crop destroying pests, whether ground squirrels, gopher, blackbird,

crows, buzzard, rats, or grasshopper, is by clouds of gas."[303] Chemical warfare metamorphosed into pest control, with the preservation of humanity, rather than its destruction, as the goal. At the same time, pest control research justified the continued existence of the Chemical Warfare Service and its improvement of poison gases.

These efforts on the part of the Chemical Warfare Service coincided nicely with the needs of civilian entomologists, who at the time worked in a low-status field considered inconsequential by society, but who played an outsized role in pest control for the war effort. Following the war, entomologists "were surprised and chagrined to find that even in certain high official circles the old idea of the entomologist still held—that he was a man whose life was devoted to the differentiation of species by the examination of the number of spines on the legs and the number of spots on the wings."[322]

By framing their work as a war against insects with human survival at stake, entomologists elevated their prestige while they eagerly embraced the tools of war: airplanes, poison gases, and dispersal weapons.[303] The Chemical Warfare Service provided them with these tools and with the fruits of its research, and hence the army and entomologists each promoted the public image of the other. The strategy succeeded with the passage of the National Defense Act in 1920, which solidified the position of the Chemical Warfare Service within the US Army.

Human progress depended, according to a typical popular article published in 1915, on mankind defeating "germ-conveying agents . . . whose only purpose in life seems to be to play the part of the anarchist and to reduce the living world to nullity and death.— There is a war to be waged, not between man and man, but between man on the one side and the arthropod on the other, a war to be fought to the finish to decide which of the two forms of life, this highly developed vertebrate or these malignly evolved invertebrates, is to govern our planet. Is the lord of this earth some day to be a monstrous ant or bug, a wasp or a midge, a scale insect or a tick? Or is it to be this god-like mammal that walks erect and can see the stars, can weigh the suns and planets, that is already in touch with

the supramundane universe? Well, the outcome of the struggle is almost as much a toss-up at the present moment as is the result of this devastating war."[328]

America's most prominent entomologist, Stephen A. Forbes, wrote, "The struggle between man and insects began long before the dawn of civilization, has continued without cessation to the present time, and will continue, no doubt, as long as the human race endures . . . We commonly think of ourselves as the lords and conquerors of nature, but insects had thoroughly mastered the world and taken full possession of it long before man began the attempt . . . It is like a war between two nations, one of which should so greatly excel in the construction of its firearms and the other in the quality of its ammunition that neither could ever gain a decisive and final victory."[329] In 1915, referring to the San Jose scale, a destructive insect that had arrived in the United States from Japan in 1872, Forbes called it "a case of Japanese invasion far more successful, and probably more destructive also, than any which Japan could possibly make by means of dreadnoughts and armies of little brown men."[329]

The Chemical Warfare Service slipped into the new niche created by military-industrial cooperation and changing public perceptions after the war. Its projects for the public good ranged widely: chemicals to defend marine pilings from corrosion, gas mask development for mine workers, rat eradication in collaboration with the Public Health Service, and earthworm and gopher extermination using poison gases.[303] The Chemical Warfare Service even jumped into the medical arena by treating people suffering from colds, bronchitis, and whooping cough with chlorine gas inhalations. The public response was heartening. "Fumes that broke the British front," read a news headline, "leaving hundreds dying, now cure influenza, bronchitis and other ailments."[303] Twenty-three senators, 146 representatives, and even President Calvin Coolidge were among the hundreds of people gassed in the US capitol for respiratory ailments.

Chemical companies, which had built substantial infrastructure and expertise during the war, capitalized on these assets to pro-

duce pesticides—and especially pursued the promise of synthetic organic compounds effective at killing insects.[303] These insecticidal efforts bolstered the public image of the chemical companies, which were regularly accused of profiteering from war and the production of poison gases. Yet, at the same time, pesticide producers had to contend with public fear of their products. One chemical executive explained that "the housewife—if she knew there was such a thing as an insecticide—bought a package in fear and trembling lest its contents kill her and her family."[303]

A toxic public image of another kind now plagued Haber. For Haber, the cessation of hostilities meant fear of being brought to trial for war crimes.[301] He had reason to worry. His was one of the names on a list considered for prosecution by the Allies. He grew a beard to disguise himself and left Germany for Switzerland. Nernst also found his name on the list of potential war criminals.[295] This infuriated Nernst, and he blamed Haber. Nernst sold his estate to convert it to liquid assets and fled the country using a fake passport provided by the German Ministry of Foreign Affairs. But in the end only a handful of Germans were tried, and they received light sentences.[301] Both Haber and Nernst returned to Germany.

Despite the waning of the war crimes threat, Haber's image was indelibly linked to the horrors of poison gases. The Haber-Bosch synthetic nitrate plant at Oppau exploded in 1921, killing more than six hundred employees and injuring more than two thousand.[326] The *New York Times* speculated that the explosion was the result of "covert experimenting by those chemists," including Haber. The journal *Nature* published its opinion on Haber in 1922: "It will not be forgotten that it was at the Kaiser Wilhelm Institute for the Promotion of Science that *Geheimrat* Haber made his experiments on poison gas, prior to the Battle of Ypres, which initiated a mode of warfare which is to the everlasting discredit of the German."[301]

Both sides, of course, claimed unfair tactics used by the other. The Germans even claimed that the Allied use of colonial troops was a cruel stratagem. The chief of the US Chemical Warfare Service shot down this claim, writing in a typical racist viewpoint of the time:

All of us recall that early in the war the Germans spread broadcast charges that the Allies were using unfair and inhumane methods of fighting because they brought the Ghurka with his terrible knife from Asia and the Moroccan from Africa. And we all know that after a time the Germans ceased saying anything about these troops. What was the cause? They were not efficient. Just as the Negro will follow a white officer over the top in daylight and fight with as much energy and courage and many times as much efficiency as the white man, he cannot stand the terrors of the night, and the same was true of the Ghurka and the Moroccan. All the Allies soon recognized that fact as shown by their drawing those troops almost entirely away from the fighting lines. In some cases dark-skinned troops were kept only as shock troops to be replaced by the more highly developed Caucasian when the line had to be held for days under the deadly fire of the counter attack. The German idea, and our own idea prior to the World War, was that semi-savages could stand the rigors and terrors of war better than the highly sensitive white man. War proved that to be utterly false.[297]

The Allied powers demanded seemingly impossible reparations from Germany, the astronomical sum of $33 billion.[301] The amount could not be gained through Haber's nitrogen fixation process because Allied nations were producing their own ammonia through the same procedure. Similarly, the dye industry, which Germany had dominated before the war, was now controlled by emergent firms in the United States. This was in large part due to the seizure of forty-five hundred German dye and chemical patents by the US government under terms of reparations.[317] These patents were then licensed to American chemical companies. One of these formerly German patents was for Haber's nitrogen fixation process. Furthermore, Germany had lost its colonies and whatever wealth could have been squeezed from them. Haber calculated that the amount of the reparations was the equivalent of 50,000 tons of gold.[301]

The Swedish scientist Svante Arrhenius suggested a solution to Haber—that he extract gold from the oceans to pay the German reparations.[301] Arrhenius, like Haber and Nernst, possessed one of the greatest minds of his generation. He had correctly postulated in his 1884 thesis the radical idea that when salt was dissolved in water, it spontaneously dissociated into positive and negative ions—Faraday's cations and anions—and that an electric current was unnecessary to this process.[295] Therefore, the forces underlying chemical binding were electric. Arrhenius's professors did not understand the significance of his thesis and judged it "not without merit," which was a grade that should have precluded an academic career.[295] But Arrhenius continued to make important discoveries on the chemistry of ions, as well as on such disparate phenomena as the greenhouse effect,[330] the northern lights, and the relationship between toxins and antitoxins.[331,332] For his work on ions, Arrhenius received the 1903 Nobel Prize in Chemistry.

Arrhenius influenced the careers of both Haber and Nernst. Early in their careers, Arrhenius convinced Nernst to work with him under the guidance of Wilhelm Ostwald, who received the 1909 Nobel Prize in Chemistry for his discoveries on catalysts.[331] Catalysts promote chemical reactions without any change in their own nature, and Haber's brilliance at finding just the right catalyst for a particular reaction allowed him to solve the ammonia synthesis problem.

Following up on Arrhenius's suggestion, Haber calculated that the oceans contain 8 billion tons of gold.[301] The problem was how to remove it from water. The proposition seemed ludicrous, but Haber had already mined wealth from the air, so why not from the oceans?

Haber constructed an analytical laboratory and gold extraction apparatus on a passenger ship of the Hamburg-American Line.[301] The effort was top secret, and was soon replicated, with his students in charge, on various other passenger ships. Reporters and bystanders speculated that Haber was generating electricity from the sea, or studying corrosion, or investigating the force necessary to stop a ship in motion, or determining how to color water. Haber reasoned

that the oceans in different parts of the world would contain varying amounts of gold, so he had ten thousand bottles of marine water shipped to him from every corner of the Earth.

After working on the problem from 1920 to 1928, Haber concluded that although gold could be extracted from sea water, the process was not economically feasible.[301,305] "The prospect of relieving Germany's burden of tribute payments," wrote Haber's friend Richard Willstätter, "turned out to be a mirage."[305] Deeply disappointed, Haber realized that he could not save Germany from its fiscal crisis, nor could he salvage his own scientific reputation, with another revolutionary innovation. He asked Willstätter if he should continue with the gold extraction research. Willstätter replied, "If it doesn't produce any gold, it will make a nice book."[305] Haber never wrote that book. He seemed to need to rely on his credentials and achievements rather than the promise of future discovery. He carried a business card that read, "Professor, Dr. Phil., Dr. Ing., E. H., Dr. d. Landw., E. H., Fritz Haber, Nobel Prize winner, former Director of the German Chemical Warfare Service, Director of the Kaiser Wilhelm Institute for Physical Chemistry and Electrochemistry."[301]

A spinoff of Haber's institute achieved economic success in the production of insecticides derived from hydrocyanic acid, which had been used as an insecticide since the late nineteenth century and was a major component of war gases. Hydrocyanic acid is lighter than air, which posed a problem for its unaltered deployment on the battlefield, but also led to the discovery of hydrogen gas as an effective mode of lift for Zeppelin airships.[333]

When Nicolle discovered in 1909 that body lice transmit typhus, the stage was set for widespread lice eradication just in time for the inevitable epidemics associated with World War I. The preferred methods of disinfestation of enclosed spaces were sulfur dioxide and steam.[333] Sulfur dioxide had the distinct disadvantage of being flammable and explosive. Steam often damaged personal property.

In 1917, German chemists tested the efficacy of hydrocyanic acid against different insect pests and found it to kill lice and their eggs (nits) as well as bedbugs.[333] Experiments were conducted to

determine the ideal strength and application temperature. They demonstrated that hydrocyanic acid outperformed both sulfur dioxide and steam. Furthermore, it was inexpensive, did not damage personal property, penetrated even tiny folds in clothing or bedding, and did not risk fire or explosion. Hydrocyanic acid was immediately used in high-risk areas for typhus outbreaks, such as military hospitals and barracks, and thereby displaced sulfur dioxide and steam. The Germans found efficient ways to seal rooms to create gas chambers into which hydrocyanic acid could be introduced to completely eradicate lice.

That same year, the German government created the Technische Ausschuss für Schädlings-bekämpfung (Technical Committee for Pest Control, called "Tasch") and appointed Haber as its chairman.[334] Tasch's mission was "to preserve through the application of highly poisonous substances for pest control, the high output that has hitherto been maintained in agriculture and forestry, in viticulture, horticulture, and fruit growing, as well as in industry by the destruction of animal pests; to promote human and animal hygiene; and thereby to ward off diseases."[334] The German military established a pest control battalion to apply the hydrocyanic acid in important food-processing facilities, barracks, hospitals, and military prison camps. Between 1917 and 1920, Tasch disinfected 21 million cubic meters of buildings with hydrocyanic acid.[333]

Clearly, the compound held profit potential, and Haber facilitated its commercialization into a private enterprise.[334] This had been his intent since the final day of World War I, November 11, 1918, when Haber told officers at the War Ministry that his goal was "to turn the means of extermination into sources of new prosperity."[312] Tasch morphed into the Deutsche Gesellschaft für Schädlingsbekämpfung (German Corporation for Pest Control, called "Degesch"). Degesch's statutes specified that "the object of the company's existence is the control of animal and plant pests using chemical means . . . The company is permitted to carry out all transactions suitable for promoting the company's purpose."[334] The intentions of Degesch fell squarely on the side of society's needs. "The operation of the

company is not directed toward profit," the charter continued, "rather the company is to follow exclusively a course directed toward the public good."[334] Haber maintained a leadership role in German pest control until 1920.

The only significant shortcomings of hydrocyanic acid were the extensive training required to safely deploy the gas and the large number of accidental fatalities that resulted from improper use.[333] As a safety measure, researchers at Haber's Kaiser Wilhelm Institute for Physical Chemistry and Electrochemistry added a foul-smelling chemical to the gas mixture in 1920. In this way, the new chemical employed the opposite strategy of the old: while the designers of poison gas weapons sought to prevent any early warning of the presence of deadly gases, the designers of the insecticide aimed to provide such a warning. The researchers used a hydrocarbonic methylester as the acrid warning and named the new pesticide Cyklon.

In addition to its vastly improved safety due to the warning odor, Cyklon (renamed Zyklon, meaning "cyclone") performed as well at killing insects as hydrocyanic acid and did not spoil food.[333] The insecticide was easily moved and deployed. Best of all for the economically downtrodden Germany, Degesch controlled all production and jealously guarded the technical manuals for its use. This secrecy fell in nicely with the policies of the German government, which wanted to maintain high safety and training standards in the application of Zyklon. Applicators had to pass an examination, be deemed reliable, and must never have run afoul of the law. Prevention of typhus was a state matter, and so state legislation regulated Zyklon's use.

The quick success of Zyklon in 1920 ran headlong into its banning by the Treaty of Versailles.[333] Article 171 stated: "The use of asphyxiating, poisonous or other gases and all analogous liquids, materials or devices being prohibited, their manufacture and importation are strictly forbidden in Germany."[335] This applied to hydrocyanic acid and therefore to Zyklon.

In 1923, the Degesch chemist Bruno Tesch added chlorine and bromide as new warning odors and a chemical stabilizer to thwart unwanted chemical reactions.[333] These alterations sidestepped Ar-

ticle 171 and thereby brought Zyklon back to the marketplace. The new Zyklon received the moniker Zyklon B, and henceforward the banned Zyklon was called Zyklon A. The chlorine and bromide also enhanced Zyklon B's insecticidal power. The new formulation was packaged more conveniently as pellets in a sealed can rather than the previous liquid-gas combination. Bruno Tesch established a company in Hamburg to distribute Zyklon B.

Zyklon B became a profitable and much needed export product for German industry.[333] Disinfestation occurred around the world in barracks, ships, trains—any enclosed space at risk for an influx of lice. Haber helped Degesch establish overseas markets, and before long the company employed agents in thirty-nine countries.[334]

In Germany, cities established gas chambers where people could delouse their clothing and furniture.[333] For the convenience of the populace, gas chambers were also driven to communities on the backs of trucks. Applicators were trained under strict governmental and industrial rules of secrecy, and tight bonds formed between the manufacturers, public health personnel, and government regulators. The meticulous care underlying the use of Zyklon B for the good of public health subsequently mutated into its monstrous employment for genocide under Nazi rule.

Zyklon B (1922–47)

It is sad to think that he spent the last months of his distinguished life an exile from his native land, finding a friendly welcome in England, which sixteen years ago was filled with feelings of great hostility toward Germany and Haber. We are not fully acquainted with the cause of his exile but believe it to be due to the descent of his ancestors from that Semitic race from whom we Europeans learned 3,000 years ago our alphabet and 2,000 years ago our religion.—**Obituary for Fritz Haber published in the British magazine *Chemistry & Industry*, 1934**[301]

Adolf Hitler, who had been temporarily blinded near the end of World War I by a British mustard gas attack, admired the wartime

delousing efficiency of German entomologists applying hydrocyanic acid.[333] When Hitler and the Nazis came to power, they found that the state and industrial controls of Zyklon B established by the Weimar government perfectly fit the Nazi Party's desire to control its use. The Nazis established a strict delousing program in both military and civilian areas based on Zyklon B.

The rise of Nazism was abhorrent to Haber, who was devoutly nationalistic for the values of the Germany of his youth—the Germany that valued the highest standards of education and led the world in research productivity.[301] When Haber heard Adolf Hitler give a speech on the radio in 1932, he said, "Give me a gun so that I can shoot him."[301] It seemed incomprehensible to him that the nation that produced the world's greatest scientists also fostered Nazism. But the postwar period was one of great instability in Germany. In 1922 alone, assassins took the lives of more than three hundred leading German Republicans.

One of those assassinated in 1922 was Walther Rathenau, the son of the industrialist who had purchased Nernst's lightbulb patent.[295] During World War I, the Kaiser appointed Rathenau to coordinate German industrial activities to support the military. A close friend of Nernst's, Rathenau advanced to become Germany's foreign minister after the war and worked relentlessly for a peaceful Europe. Attackers riddled his car with bullets from a machine gun while they tossed in a hand grenade. The assassination devastated Nernst.

Meanwhile, Ludendorff, who fled to Sweden at the end of the war, proclaimed that the German army had been on the verge of victory when it was "stabbed in the back by the Jews."[295] Ludendorff had been raised a Christian, but he left that faith behind because its founder was a Jew. The Nazis rallied around Ludendorff's excuse for Germany's defeat. Anonymous threats ensued, calling for the assassination of the "relativity Jew" Einstein. Ludendorff returned to Germany and joined the Parliament as a member of a coalition party made up of members of the German Völkisch Freedom Party and the Nazi Party. He mounted an unsuccessful campaign against Hindenburg for the German presidency.

Runaway inflation reached a point where a newspaper cost more than 100 billion marks. One US dollar traded for 4 trillion marks. One German mark in 1918 had the same value as 500 billion marks in 1923.[326] The ranks of the unemployed topped 7 million.[295] The strength of the Nazi Party grew along with Germany's discontent.

On January 22, 1933, Hitler had a long private conversation with Hindenburg's son at a dinner party.[295] The Nazis had acquired information proving corruption on the part of the Hindenburg family, and Hitler used it for blackmail. On January 30, Hindenburg asked Hitler to become chancellor and form a government. Within two months the Nazi Party controlled all public meetings, suspended freedom of the press and civil rights guaranteed by the German constitution, arrested communists and liberals, appropriated complete police power, and solidified its dictatorship with the Enabling Act.[301]

The 1933 Aryan clause of the Civil Service Law of the Third Reich stripped German Jews, except war veterans, of their government employment. This early Nazi attack on Jews had a disproportionate effect on German science.[301] Although Jews composed less than 1 percent of Germany's population, one-third of Germany's Nobel Prize–winning scientists were Jewish. Nazi university students threw their weight behind the party by creating an antisemitic declaration that libeled Jewish professors. Albert Einstein, at the time giving lectures at Princeton University, returned to Europe to tender his resignation from the Prussian Academy of Sciences.[308]

Two German Nobel Prize winners in physics, Philipp Lenard and Johannes Stark, sided with the Nazis and attacked Einstein as a traitor.[295] Jews, they argued, poisoned German physics. Both men had been passed over for the professorship in Berlin that had been given to Nernst in 1924. Now they could take advantage of the dismissal of prominent Jews in academia. Stark replaced a dismissed Jewish scientist as the president of an important physics institute. Lenard was rewarded for his loyalty to the Nazis with the post of chief of "Aryan Physics."[336] He wrote a four-volume work on German physics, liberated from "Jewish physics, a degenerate mirage of

the basic Aryan physics."[295] Its publication was celebrated by a Nazi student group, who lauded Lenard's courage for his lecture in which he celebrated the assassination of Rathenau.

An unknown German physicist, Wilhelm Müller, published a book stating that an international Jewish conspiracy sought to corrupt science and destroy humanity.[295] He was then appointed professor of theoretical physics at Munich, where he replaced the retiring professor and world-renowned physicist Arnold Sommerfeld.[337] Müller's appointment was celebrated with a lecture by Stark on "Jewish and German physics."[295] The fact that the Nazis would replace Sommerfeld, who had trained more Nobel Prize winners than any other scientist, with a fraud as the head of a major physics institute reflected their disregard for scientific discovery. That disregard manifested itself in many ways, including the Nazis' subsequent failure to recognize nuclear fission even though its discovery had occurred in a German laboratory.

Haber's friend, protégé, and the 1925 winner of the Nobel Prize in Physics, James Franck, also resigned his professorship and institute directorship in protest of the Nazi policies.[308] Franck explained his decision to journalists with the comment, "How can I teach students or examine them if they think about me and my forefathers like that?"[301] Franck wrote in a letter to Haber, "Don't scold me as impetuous or thoughtless. I expressed my wish to the Minister today that I be released from my duties. The reason I gave was 'the attitude of the government toward German Jewry.' . . . I simply cannot stand before the students at the beginning of the semester, whose representatives have set up the precepts you know about, and act as if none of all this concerned me. Nor can I gnaw at the bone of clemency that the government holds out to war veterans of Jewish race."[308]

Haber felt deeply disappointed that Franck would not be his successor as director of the Kaiser Wilhelm Institute for Physical Chemistry and Electrochemistry, a move that had been imminent.[308] Earlier, he had also planned for Franck to succeed Nernst in his institute directorship. He wrote in a letter to Franck, "The

thought that this institute should not fall into your hands is the bitterest drop in the brew of these weeks. You wish to stay in Germany and I, although not having the wish, do not see the possibility of proceeding otherwise. I would not know how to emigrate honorably and find an existence abroad in my later years."[308]

Haber tried all his life to hide from his Jewish roots.[301] He had converted to Christianity and impressed upon his Jewish students that they should do the same. Although he thought a Jewish homeland in Palestine might be fitting for Eastern European Jews, he felt that it certainly was not for German Jews. Yet both of Haber's wives were Jewish, most of his friends were Jewish, and Nazi Germany considered him a Jew no matter what his self-perception.

Haber was initially exempt from dismissal due to his standing, but he resigned his directorship and full professorship in protest. In his resignation letter he wrote, "According to the provisions of the Government Employees Law of April 7, 1933, which were ordered to be applied to the institutes of the Kaiser Wilhelm Society, I am entitled to remain in office although I am descended from Jewish grandparents and parents. However, I do not wish to take advantage of this permission any longer than is necessary for the orderly discharge of the scientific and administrative duties of my offices."[305] Haber continued, "I tender my resignation with the same pride with which I have served my country during my lifetime . . . For more than forty years I have selected my collaborators on the basis of their intelligence and their character and not on the basis of their grandmothers, and I am not willing for the rest of my life to change this method which I have found so good."[301]

At the same time, Nernst arrived at work and sought out a Jewish colleague.[295] When he learned that the colleague had been denied entry into the laboratory because he was a Jew, Nernst became enraged. Nernst proceeded straight to Haber's institute where he told Haber that conditions in his own institute were intolerable, and he asked Haber for a job. Haber was, at that moment, packing up his own office, having tendered his resignation, and he was in no position to help Nernst. Clearly, the rivalry that had preoccupied

the men in their simultaneous rise to prominence seemed pedestrian compared to the harsh reality of Nazi policies.

Meanwhile, the Nazis purged Jewish scientists from the leadership of the national institutes. One Nazi memo read: "The existing board of trustees of the Physikalisch-Technische Reichsantalt (P.T.R.) still has its old composition from before the rise to might of National Socialism. The traitor Einstein has been excluded but there still are Jews and notables of the old system among the remaining members . . . On the board today there are people like Professor James Franck (full-blooded Jew); Professor Haber (full-blooded Jew); Professor Hertz (half-Jew); . . . and Professor Nernst, one of the strongest advocates of the liberalist and capitalist world view."[308] Although he was not Jewish, the 1933 Nobel Prize winner Erwin Schrödinger resigned his professorship and left Germany that same year in protest of the Nazi treatment of Jewish colleagues.[338]

Franck accepted an invitation from the eminent Danish physicist Niels Bohr to collaborate in his institute in Copenhagen, and subsequently took a professorship in the United States.[308] He wrote in a letter to a German colleague, "It appears that there is no room left for people of my heredity and my temperament in Germany to work with success . . . You know that I feel just as good a German as any other, but that won't help me; I simply have to emigrate, even though I know that my wife and I won't be able to grow roots anywhere else; but perhaps the children will succeed, or at least the grandchildren. I do not want them to have to feel like second-class citizens."[308]

Franck and his fellow German physicist Max von Laue both left their Nobel medals, which were made of gold, with Bohr for safekeeping from the Nazis.[339] When the Nazis invaded Denmark, and just before they appropriated Bohr's institute, Bohr felt preoccupied with how to keep the medals out of their hands. Bohr had already donated his own Nobel medal to a Finnish relief fund, so he was not concerned about that. While Franck was by that time in the United States, von Laue remained in Germany and engaged in anti-Nazi activities including standing quite alone in his protests against

persecution of Jewish colleagues.[295] The discovery of exported gold with his name on it would have meant certain arrest.

A Hungarian chemist in Bohr's institute, and future Nobel Laureate, George de Hevesy, intervened. "I suggested that we should bury the medal," wrote de Hevesy, "but Bohr did not like this idea as the medal might be unearthed. I decided to dissolve it. While the invading forces marched in the streets of Copenhagen, I was busy dissolving Laue's and also James Franck's medals."[340] De Hevesy dissolved the two medals together in nitro-hydrochloric acid—a difficult prospect given the limited reactivity of gold. He put the solution on a laboratory shelf, where it remained untouched and unnoticed by the Nazis throughout the war. After the war, the gold was precipitated from the solution and dispatched to the Nobel Society, where the two medals were recast and given back to Franck and von Laue. Theirs are the only amalgamated Nobel medals.

Eventually the situation in Denmark became untenable for Bohr.[341] Bohr had helped many prominent Jewish and dissident scientists escape from Nazi Germany. Additionally, Bohr's own mother was Jewish. British Intelligence discovered that his arrest was planned, and so they sent a message to him on microfilm hidden inside two rusted keys. The British did not want the Nazis to incarcerate a man who could accelerate a German nuclear weapons program, and they hoped to acquire Bohr's expertise for their own nuclear program. Bohr, in his turn, did not want to abandon his country or the colleagues he continued to help, and he stayed until he received a message from an anti-Nazi in the German government that his arrest was imminent.

With the help of the Danish resistance, Bohr and his wife slipped away from their Gestapo followers and crossed the sea to Sweden in a fishing boat.[341] During the next few weeks, the Danish resistance shuttled almost the entire Jewish population of Denmark in small boats at night across the sea to Sweden, in the only case of a Nazi-occupied country saving its Jewish population. Included in the armada of refugee boats were Bohr's children. The wife of a Swedish embassy official carried one of Bohr's grandchildren to Sweden in a

shopping basket. Meanwhile, Bohr held secret meetings with Sweden's king and politicians to facilitate the influx of refugees.

With his family safely together again in Stockholm, Bohr accepted an invitation by the British government to help the British with the war effort.[341] On October 6, 1943, a British mosquito bomber, containing no bombs because of Sweden's neutrality, flew Bohr from Stockholm over Nazi-held Norway to the United Kingdom. To avoid Nazi anti-aircraft guns, the pilot flew at great altitude. Given the limited space in the plane, Bohr lay in the bomb bay in a flight suit with a parachute strapped on. Unfortunately, Bohr's head was so large that he could not pull on the provided helmet completely and so he did not hear the pilot's instruction via the earphones to turn on his oxygen supply. Bohr passed out from lack of oxygen, only regaining consciousness when the plane descended over the sea past Norway. Operating under the utmost secrecy, Bohr then worked with Franck and others on the British and American nuclear weapons program.

Shortly after the Nazis seized power, Haber joined the exodus of Jewish scientists from Germany.[301] While on a trip to Switzerland, he visited with the biochemist and Zionist leader Chaim Weizmann, who would later become the first president of Israel. Weizmann had previously offered Haber (and James Franck)[314] a professorship in Palestine, but Haber declined as he did not approve of Zionism. As the Nazis consolidated their power, Haber reconsidered his views. Haber said, "Dr. Weizmann, I was one of the mightiest men in Germany. I was more than a great army commander, more than a captain of industry. I was the founder of industries; my work was essential for the economic and military expansion of Germany. All doors were open for me. But the position I occupied then, glamorous as it may have seemed, is as nothing compared with yours. You are not creating out of plenty—you are creating out of nothing, in a land which lacks everything; you are trying to restore a derelict people to a sense of dignity. And you are, I think, succeeding. At the end of my life I find myself a bankrupt. When I am gone and forgotten, your work will stand, a shining monument, in the long history of

our people."[334] Haber then agreed to take the position in Palestine, and to come there with his sister. But first he accepted an invitation to work in William Pope's laboratory at Cambridge University.

Pope's invitation to Haber reflected forgiveness, their intellectual connection through science, and their shared hatred of the Nazis. Pope had developed mustard gas for the Allies in World War I to counter the war gases developed by Haber, and English feelings toward Haber had been extremely bitter since the end of the war.[301] After a short visit to Pope's lab, at the end of January 1934, Haber returned with his sister to Switzerland for a three-day visit on his way to his new position in Palestine. He died in his sleep on the third day.

What led to his death, according to Willstätter, was not overwork: "Great tasks and efforts stimulated and energized Haber, as they do other strong personalities. Neither during the war nor during preceding decades, nor in the last years of Haber's life was it overwork that undermined his health and caused his premature death. It was rather his terrible disappointment at the outcome of the war and the conditions of the peace, and at being considered an outlaw by the Allies. Then as always it was the poisons of sorrow, pain, and suffering that consume a man and exhaust him."[305]

Von Laue, who had offered Haber's students places in his lab, wrote an obituary that was published in Germany, and both he and his editor were severely reprimanded for it by the Nazis.[295] For the one-year anniversary of his death, the Kaiser Wilhelm Society, the German Chemical Society, and the German Physical Society held a memorial ceremony attended by more than five hundred people. The Reich minister for science, education, and national culture banned the event with a circular that read: "Professor Haber was dismissed from his office on October 1, 1933 in accordance with a request from him which expressed unequivocally his inner opposition to the present government and which the public could not fail to regard as criticism of the measures instituted by the National Socialist State. The intention of the above-mentioned societies to hold a memorial service on the occasion of the first anniversary of his death must, therefore, in the face of these facts be construed as

a challenge to the National Socialist State, since such an anniversary is accorded special recognition only in very exceptional cases, with the greatest Germans."[305] This small nod to Haber's legacy constituted the only organized academic resistance to Nazism from within Germany.[295] "It was a large, select assemblage that filled the auditorium," wrote Willstätter, "but the nonparticipants were more noticeable than the participants. Every participant was required to identify himself and enter his name on a list."[305]

The parallels between Haber and Nernst ran throughout their lives, all the way to the bitter end under Nazi rule. Like Einstein, Franck, and Haber, Nernst resigned his professorship in 1933.[295] Nernst had already paid dearly for the belligerence of Germany. Both of his sons had died in combat in World War I. By the time the Nazis rose to power, two of Nernst's three daughters were married to Jewish men, and they fled abroad with their children. The last of Nernst's grandchildren escaped from Germany in 1939. That year Nernst suffered a heart attack. He died two years later. His ashes were subsequently moved to rest next to those of the eminent physicist Max Planck and von Laue in Göttingen. Nernst had hoped to be able to support his grandchildren long after he was gone, so his last major financial decision was to convert his considerable assets and cash into woodlands that would yield timber for generations. When Nazi Germany fell, the woodlands became part of Poland.

When Germany invaded Poland in 1939, German authorities grew concerned about a potential typhus outbreak.[333] They quickly employed Zyklon B gas chambers, replicating the successful German World War I program with hydrocyanic acid. Meanwhile, the Nazis struggled to find a way to efficiently kill the millions of Jews they were rounding up into concentration camps.

The first experimental gas chamber for exterminating people became operational in 1939.[342] The gas of choice at that time was carbon monoxide. In the first experiment, Hitler's personal physician Karl Brandt, the head of the Nazi Euthanasia Program Philipp Bouhler, and the liaison with the Health Department Victor Brack were among the officials who observed the gassing of four men-

tal patients. Gas chambers, disguised as shower rooms, were then constructed for multiple hospitals to eliminate the insane. Among those killed by the Nazis were German veterans of gas warfare from World War I, whose mental health had never recovered following exposure to chemical weapons.[306] Thousands of Jews and other concentration camp victims were also brought to these hospital gas chambers to be killed, but these gas chambers did not have the capacity that the Nazis desired.[342]

In the summer of 1941, the annihilation camps were constructed. Heinrich Himmler, who had headed the SS (Schutzstaffel, Nazi paramilitary organization) since 1929 and held the title of "Commissioner for the Improvement of German Folkdom and the Trustee of Race Purity," instructed Rudolf Höss, the commander of Auschwitz, to implement efficient methods for the "final solution of the Jewish question."[333,342,343] Himmler said to Höss, "The Jews are the eternal enemies of the German people and must be exterminated."[343]

Höss found the gas chamber at Treblinka, which used carbon monoxide, to be wanting in speed and lethality.[333] Adolf Eichmann, in charge of the Final Solution, agreed. Eichmann discussed his concerns about the various methods of killing Jews with Höss on a visit to Auschwitz.[343] Part of the problem was the effect that gunning down women and children had on some of the SS men. But the bigger issue was the practicality of killing millions of Jews by the existing methods of carbon monoxide and shooting, which were both too cumbersome and too slow.

The administrative chief of Auschwitz, Karl Fritsch, believed that Zyklon B could solve their problem.[343] Fritsch tested its killing power on prisoners and proved its effectiveness. Then, as a proof of concept for mass executions, in September 1941, Höss's team used a Zyklon B disinfestation gas chamber to kill six hundred Russian prisoners of war, including many Jews, and two hundred fifty mental patients.[342] Höss, after further experimentation on concentration camp prisoners, concluded that Zyklon B killed far more efficiently than carbon monoxide.[333] The warning odor of Zyklon B was removed in the SS's modification for human extermination.[344]

Officials at Degesch initially opposed this order to remove the warning odor from Zyklon B because their patent on Zyklon B had expired, but the patent on the warning odor was still in effect.[326] Degesch's monopoly on the production of Zyklon B depended entirely on the inclusion of this chemical indicator. But Degesch complied with the SS demands and removed the indicator in the new version of Zyklon B destined for the extermination camps.

Gas chambers were constructed on farmland selected by Höss and Eichmann, and inscribed with the words *"Zur Desinfektion."*[333,343] The new gas chambers constructed in Auschwitz's killing center at Birkenau allowed ten thousand people to be killed each day. Jews and other victims, told they would be deloused, were forced to undress and then marched between lines of police wielding whips, sticks, and guns into the gas chambers, where they died after three to twenty minutes of exposure.[333,342,343] Höss admired this efficiency. At Treblinka, ten times as many gas chambers were needed to kill Jews at that pace with carbon monoxide.

Carbon monoxide cost more than Zyklon B, took longer to kill, and was more difficult to handle.[333] Its use had started with killing mental patients, then it was adapted for moveable gas chambers, and then it was implemented on a large scale in death camps. But the employment of Zyklon B made carbon monoxide obsolete.

The only difficult requirement for Zyklon B was the maintenance of a temperature of 25.7°C, to exceed its boiling point of 25.6°C and hence volatize the gas.[333] To accommodate this, the SS allowed the body heat of the victims to warm the gas chamber to the correct temperature before dropping the tins of Zyklon B through holes in the roof. The SS men were trained in the safe handling of Zyklon B and wore gas masks throughout the operation. Afterward, ventilators removed the gas, and then Jewish prisoners, the Sonderkommando, were forced to strip the corpses of everything of value, such as hair and teeth with gold fillings, before they incinerated the bodies.[343]

Between 1942 and 1945, the Nazis killed more than 5 million people in Auschwitz, Bełżec, Chełmno, Majdanek, Sobibór and Treblinka, mostly with Zyklon B.[342] After the war, Höss stated to the

War Crimes Investigation Unit of the British Army of the Rhine, "I personally arranged on orders received from Himmler in May 1941 the gassing of two million persons between June–July 1941 and the end of 1943, during which time I was commandant of Auschwitz."[342]

World War II far surpassed World War I in destruction and resulted in more than half of the war fatalities that had occurred over the past two thousand years.[303] A large fraction of these fatalities were civilians in cities that were hit with firebombs—that is, chemical incendiary weapons. Another large fraction were the 6 million Jews killed in the concentration camps, death marches and ghettoes. When Eichmann reported to Himmler that they had succeeded in killing 6 million Jews, Himmler expressed his disappointment with such a low figure.[343]

Bruno Tesch owned the company Tesch & Stabenow, which supplied Zyklon B to Auschwitz/Birkenau.[345] Tesch and his deputy Karl Weinbacher were tried by the British Military Court in Hamburg in March 1946 for war crimes under Article 46 of the Hague Regulations of 1907. Article 46 stated: "Family honour and rights, the lives of persons, and private property, as well as religious convictions and practice, must be respected. Private property cannot be confiscated."[346] Germany and Great Britain were both parties to the Hague Regulations and Great Britain was the ruling authority of that sector of Germany at the conclusion of the war. According to the charge, Tesch and Weinbacher "at Hamburg, Germany, between 1st January, 1941, and 31st March, 1945, in violation of the laws and usages of war did supply poison gas used for the extermination of allied nationals interned in concentration camps well knowing that the said gas was to be so used."[345]

Wehrmacht leaders had told Tesch early in the Holocaust that shooting and burying Jews in large numbers was unhygienic, and had asked him what he thought about using Zyklon B.[345] Tesch agreed that this would be a better method, and could be done in a gas chamber in the same way that vermin were exterminated. Tesch and his firm then supplied the Zyklon B and provided expert technicians and training for the Wehrmacht and the SS.

For a conviction, the court had to be sure of three facts: "first, that Allied nationals had been gassed by means of Zyklon B; secondly, that this gas had been supplied by Tesch and Stabenow; and thirdly, that the accused knew that the gas was to be used for the purpose of killing human beings."[345] Among many pieces of evidence considered by the court was an invoice from Tesch & Stabenow that read: "We have sent to Auschwitz as freight goods, the following shipment of Zyklon B cyanide, without an irritant."[343] This, of course, was the modification of Zyklon B for human extermination, since German law stipulated that Zyklon B for insecticidal use must contain the warning odor for the safety of personnel. Tesch and Weinbacher were found guilty by the court and hanged.[345]

As Germany crumbled at the end of the war, and after Hitler took his own life, Höss reported to Himmler in Flensburg.[343] Himmler's order was, "Plunge out of sight into the ranks of the army."[343] Höss donned a naval uniform and morphed into boatswain Franz Lang, in whose identity he passed through British inspections. Devoted to his wife and children, whom he lavished with attention during their residency at Auschwitz, Höss worked on a farm near to them for eight months. Meanwhile, the British military police hunted for him in vain. The British Field Security Police finally caught Höss on March 11, 1946. With him, packed alongside his clothing and other personal effects, was his Auschwitz horsewhip. The man who ordered and witnessed the daily massacre of ten thousand Jews could not part with his whip, a symbol of his power over others, even though its presence might give him away.

At Nuremberg, unlike most of the other defendants, such as Tesch and Weinbacher, who proclaimed their innocence, Höss elaborated upon his crimes.[343] Höss's only expression of remorse came on April 12, 1947, four days before the court hanged him on the gallows in Auschwitz, when he wrote, "In the isolation of my imprisonment, I came to the bitter recognition of how deeply I transgressed against humanity."[343] Yet, in Höss's mind, his transgression was against the Polish people, of whom he begged forgiveness; he showed no remorse for his destruction of Europe's Jews.

The development of Zyklon had started under Haber's watch for the purpose of improving public health in the battle against typhus. That its use morphed into the primary tool of Nazi genocide is a cruel twist of fate in the history of the creative genius who moved so readily between the research lab, commercial enterprises, and the halls of government. Haber's niece Hilde, her husband, and their two children were among the countless Jews killed at Auschwitz, probably by Zyklon B.[334]

Haber's son Hermann (who had discovered his mother Clara dying in the garden after she shot herself during World War I) escaped to the Caribbean from Nazi-occupied France with his wife and three daughters in 1942, and from there gained entry to the United States.[334] When Hermann's wife died at the end of the war, he committed suicide. Haber's second wife, divorced from Haber, moved with their son Ludwig and daughter Eva to England. The British government interned Ludwig as an enemy alien on the Isle of Man and then in Canada.[347] After the war, Ludwig became a historian of the poison gas warfare of World War I, and had himself gassed to experience the effects before writing his book on the subject.[348]

Haber's scientific redemption followed his death and the collapse of the Nazi regime when the Kaiser Wilhelm Institute for Physical Chemistry and Electrochemistry was renamed the Fritz Haber Institute.

[8]

DDT

(1939–1950)

Armed with DDT, the Army has conquered the fear of typhus. For the first time in history, this ruthless companion of disaster, famine and poverty has lost all right to its murderous title of champion of the ancient plagues of war.—**Brigadier General James Stevens Simmons**, chief of the Preventive Medicine Service, US Army, 1945[349]

The first attack of 183 combat aircraft launched from Japan's carrier fleet barraged Hawaii's Pearl Harbor just before 8 AM on December 7, 1941.[350] The second wave of 54 high-level bombers, 78 dive bombers, and 36 fighter planes attacked an hour later. The surprise attack crippled or destroyed 18 American warships and several hundred planes. American casualties included 2,400 dead and 1,178 wounded; the US Navy lost three times as many men in the space of two hours as it had during all the battles of the Spanish-American War and World War I combined.[351]

Nine hours later, Japanese bombers flying over Manila destroyed half of General Douglas MacArthur's parked fleet of B-17 Flying Fortress bombers and many P-40 fighters.[351] Japanese victories quickly mounted, with the fall of Guam (December 11), Wake Island (December 23), Hong Kong (December 25), the Dutch East Indies

(January 1942), and Singapore (February 15). After five months of conquest, Japan controlled vast territories extending to the borders of Australia and British India, and Allied forces had lost the Philippines. The Pacific War raged for four years after Pearl Harbor, with the forces of both sides stretched from the Aleutian Archipelago in Alaska south to the islands of New Guinea.

Although insects posed a mere nuisance in the North Pacific Theater, they were deadly vectors of disease in the South Pacific. General MacArthur wrote, "Millions of insects abounded everywhere. Clouds of mosquitoes, flies, leeches, chiggers, ants, fleas, and other parasites pestered man night and day. Disease was an unrelenting foe."[352]

In 1944, as Marines of the Fourth Division prepared to invade the tropical island of Saipan, their battalion surgeon briefed them:[353] "In the surf, beware of sharks, barracuda, sea snakes, anemones, razor-sharp coral, polluted waters, poison fish and giant clams that shut on a man like a bear trap. Ashore, there is leprosy, typhus, filariasis, yaws, typhoid, dengue fever, dysentery, saber grass, hordes of flies, snakes and giant lizards. Eat nothing growing on the island, don't drink its waters, and don't approach its inhabitants. Any questions?"

An astonished private asked, "Sir, why'n hell don't we let the Japs keep the island?"

Malaria was the deadliest tropical disease. A half-million American soldiers contracted malaria in the South Pacific, where the case rate for Allied troops (including recurrences) in some areas reached 4,000 per 1,000 troops per year.[63,354–56] Between October 1942 and April 1943, the ratio of Allied soldiers hospitalized with malaria to casualties on the battlefronts of the Southwest Pacific Theater was ten to one.[56] One American infantry unit fighting in New Guinea reported 2 soldiers killed in action, 13 wounded, and 925 sick.[356] Malaria among American and Filipino troops contributed to the dramatic loss at Bataan in April 1942, which constituted the greatest surrender of US forces since the American Civil War and preceded the infamous Bataan Death March.[63,357] At Guadalcanal, the division commander ordered that Marines could only relinquish their

patrol duty when their body temperature exceeded 103°F.[357] In 1943, General MacArthur complained to an army expert on malaria, "Doctor, this will be a long war if for every division I have facing the enemy I must count on a second division in hospital with malaria and a third division convalescing from this debilitating disease!"[63]

The problem stemmed largely from military commanders and soldiers who did not take the threat seriously, and the lack of command authority vested in medical officers who did.[303,356,357] This lackadaisical attitude was exemplified by an officer who said, "We are here to kill Japs and to hell with mosquitoes."[357] A critic of these problems noted, "Tough troops, accustomed to facing death from bombs, bullets, shells, and bayonets daily, they thought it sissy to make so much fuss about a tiny mosquito whose bite could hardly be felt."[358] Military activities also contributed to the outbreaks; an American military doctor pointed out that "trenches, foxholes, tank traps, gun emplacements, vehicle ruts, shell, bomb and mine craters, sabotaged irrigation projects, streams ponded by bridge rubble and improvised causeways, drainage blocked by hastily built airfields and highways, all may provide additional breeding places for malaria mosquitoes."[63]

Improving malaria control on the battlefield became a top priority, and the military employed a propaganda blitz to educate soldiers about the threat. One training pamphlet stated: "You can be made as chronic an invalid from malaria as from poison gas. Malaria can make you weak and puny, and good for nothing."[303] A clever Australian sergeant in New Guinea overcame the machismo of his troops, who refused to abide by malaria control measures, by posting signs that stated "Preserve Your Manhood! MALARIA can cause IMPOTENCE!"[358] American commanders quickly copied this tactic with great success.

For centuries, only quinine, extracted from the bark of the cinchona tree, prevented and treated malaria, and more than 90 percent of the world's cinchona supply was grown on Dutch plantations in Java.[56] Hence, in January 1942, when Japanese forces seized control of Java, they also gained control of global cinchona production. Similarly, German forces seized stockpiles of quinine in Amsterdam.

Fig. 8.1. Theodor S. Geisel (Dr. Seuss) drew a series of cartoons for the US military malaria propaganda campaign depicting the mosquito as a femme fatale.

Allied forces had access to a small grove of cinchona trees on the island of Mindanao that had been established with seeds smuggled out of Java in a malted milk bottle.[358] The Mindanao plantation had supplied quinine beginning in 1927, but this supply could barely make a dent in the demand. The fall of Mindanao in May 1942 ended any hope of accessing even that modest source of quinine.

These setbacks spurred the US government to organize an intensive research effort to discover a substitute for natural quinine.[56] This effort was not the first of its kind. A seventeen-year-old chemist named William Perkin attempted to synthesize quinine in 1856 from aniline, which is a by-product of coal tar.[358] The result was a black residue that he considered worthless, so he tried to wash the residue out of his experimental beaker with alcohol. The alcohol reacted with the residue to produce a mauve dye, and so Perkin turned his accidental discovery into a fortune and wrecked the Indian plantation business that until that time had held a monopoly on mauve

dye. Pasteur also tried but failed to produce quinine synthetically. But during World War II, American chemists successfully synthesized quinine from coal tar and tested more than fifteen thousand compounds for antimalarial effects, including numerous dyes that had been discovered by German scientists.[63]

The United States tested the efficacy of antimalarial drugs on approximately eight hundred prisoners at the US Penitentiary in Atlanta, the Illinois State Penitentiary, and the New Jersey Reformatory; these prisoners volunteered to be bitten by mosquitoes carrying relapsing vivax malaria.[359] A journalist described the patriotism of the volunteers:

> They expose themselves to further and even greater danger by taking varying doses of the new drugs to determine whether the chemicals can safely be given to our fighting men exposed to malaria and how large a dose can be tolerated by the human system . . . These one-time enemies of society appreciate to the fullest extent just how completely this is everybody's war . . . Upon learning that, through their cooperation, thousands of GI's might be spared the ravages of the tropical malady, the prisoners respond immediately and enthusiastically . . . They joke about the huge needles used for taking blood specimens, calling them "harpoons," but not one has refused to stay for the full course of punctures and tedious, frequent examinations . . . The nature of the drug or drugs, as well as the results, is still a closely guarded secret, but the stage of large-scale human testing is regarded in itself as indicating that the long-sought goal is close to realization.[359]

Apparently, this patriotic service of the prisoners had great reformative value, as indicated by the low recidivism under subsequent parole.[287]

Impelled by quinine shortages during and after World War I, the Germans had synthesized several substitutes between the world wars. The most efficacious of these was a compound called atabrine, synthesized in 1930.[63] In 1943, the United States selected atabrine

Fig. 8.2. Effective campaigns were launched on the battlefield to educate soldiers about malaria prevention. Sign posted at the 363rd Station Hospital. OHA 220.1 Museum and Medical Arts Service (MAMAS D44-145-1). From Otis Historical Archives, National Museum of Health and Medicine.

as its new quinine substitute. Also that year, Allied chemists managed to work out the secrets of atabrine synthesis and then produced it on a large scale to keep soldiers on the battlefield. But atabrine tasted bad; caused nausea, vomiting, and diarrhea; yellowed the skin (the "atabrine tan"); and led some of those who took it into psychosis.[356,357] It was also rumored to cause impotence. Many soldiers preferred to risk malaria.

Japanese soldiers used quinine prophylactics, repellents, and mosquito netting that covered an entire squad of troops at once. Yet one Japanese regiment fighting in Burma reported that all of its soldiers were infected.[355]

German army units also suffered malaria during their invasions of Greece, Ukraine, and Russia in 1941, and they, too, turned to

atabrine.[355] But German actions also caused a malaria outbreak. As American and British forces pushed against the Italian front in late 1943, the German army flooded the Pontine Marshes by sabotaging sea walls and pumping stations while also damming river and canal outlets.[360] In a matter of weeks, German forces managed to flood 100,000 acres of reclaimed farmland with a combination of salt water and fresh water. Their intention was to impede the advance of Allied troops. Ironically, the farmland had been reclaimed from wetlands in the early 1930s under an ambitious program of Hitler's ally Mussolini. A journalist reported that the "Italian campaign has been fought against three enemies—the Germans, the terrain and the mosquito."[361] She visited a farm previously occupied by Nazi soldiers and found on the pigsty walls illustrations of the plague of mosquitoes that would follow the German operation. Due to their use of atabrine, Allied forces were spared, but local civilians suffered.

The Nazis' connection to malaria also extended to their medical experiments in concentration camps. The American Military Tribunal at Dachau hanged Dr. Klaus Schilling in 1946 for war crimes. Schilling was a famous malaria research scientist and a member of the malaria commission of the League of Nations before the war, as well as the head of the Department for Tropical Diseases of the Robert Koch Institute, but he turned his skills against inmates of the Dachau concentration camp during the Holocaust.[287] Between February 1942 and April 1945, Schilling and other prominent Nazi doctors under his supervision infected more than twelve hundred inmates with malaria, either directly by mosquitoes or via injections of mosquito mucous gland extracts. The victims, including children, contracted malaria and were then subjected to experimental trials of various drugs, including quinine and atabrine. Many were given experimental immunizations and then repeatedly infected. It was through such experiments that the Germans developed their dosing guidelines for atabrine.

During the war crimes tribunal, accused doctors repeatedly invoked the American experiments on prisoners to demonstrate that

such work was routine, even though the American prisoners had volunteered for the experiments. One defense counsel argued that based on the prisoner experiment in the United States, "one must come to the conclusion that medical experiments on human beings are not only admissible on principle, but in addition, that it also does not violate the basic principles of criminal law of civilized nations to carry out experiments on convicts."[287] Schilling "believed it was his duty to humanity" to conduct these experiments in order to cure malaria. He told the war crimes tribunal "that his work was unfinished and that the court should do what it could to help him finish his experiments for the benefit of science and to rehabilitate himself."[287] A footnote to the war crimes tribunal and prosecution of Nazi doctors is that at least six of the doctors who conducted research on inmates at Dachau were secretly recruited by the US Army after the war in Operation Paperclip, in which Nazi doctors, scientists, and engineers were secretly brought to the United States to lend their expertise to the development of American science and technology.[362]

Although prophylactic chemicals prevented some strains of malaria, most tropical diseases carried by insects had no impediments. Even long-standing insecticides were unavailable. The war severed the supply line of rotenone from the Dutch East Indies, and pyrethrum stocks were depleted due to a failure of the chrysanthemum crop and labor unrest in the British colony of Kenya.[349] Furthermore, Japan had been the primary supplier of pyrethrum to the United States before the war.[363] Pyrethrum was the active ingredient of the army's louse powder and the Freon-pyrethrum mosquito bomb. None of these insecticides, even if they had been readily available, was sufficiently effective. Therefore, Allied troops were nearly defenseless against the parade of insect-transmitted diseases until 1943, when the military introduced a new weapon to the South Pacific, European, and North African Theaters: DDT, or dichlorodiphenyltrichloroethane.

The Austrian scientist Othmar Zeidler synthesized DDT in 1873–74, though he did not realize its importance and devoted to it only six lines of an article in the proceedings of the German Chemical

Society.[364,365] The Swiss chemist Paul Müller discovered DDT's insecticidal properties in 1939, and nine years later won the Nobel Prize in Medicine or Physiology for his finding.[366] That merely nine years passed between Müller's discovery and his Nobel Prize is a striking sign of DDT's immediate impact and recognition. DDT symbolized the ability of mankind to conquer any problem, even those that had plagued humanity for all time.

In 1935, Müller launched his investigation of insecticides for his employer, J. R. Geigy, in Basel. Through his review of the literature and patents, Müller found that none of the synthetic insecticides compared in effectiveness with natural insecticides used long before, including the arsenates, pyrethrum, and rotenone. "This gave me the courage to press on," he wrote. "In other respects too, the chances were worse than poor; only a particularly cheap or remarkably effective insecticide had any prospects of being used in agriculture, since the demands put upon an agricultural insecticide must necessarily be strict. I relied upon my determination and powers of observation. I considered what my ideal insecticide should look like, and the properties it should possess."[367]

Müller established seven requirements for an ideal insecticide:

1. Great insect toxicity
2. Rapid onset of toxic action
3. Little or no mammalian or plant toxicity
4. No irritant effect and no unpleasant odor
5. Wide range of action affecting as many arthropods as possible
6. Long, persistent action, which translates as good chemical stability
7. Inexpensiveness

Müller then applied these criteria to previously known insecticides, such as nicotine, rotenone, and pyrethrum, and found their effectiveness to be unsatisfactory.

Müller conducted experimental trials on the bluebottle fly, *Calliphora vomitoria*, inside a glass chamber. He tested hundreds of com-

Fig. 8.3. Paul Müller with his test apparatus, 1952. © Novartis AG.

pounds with no success. "In the field of natural science only persistence and sustained hard work will produce results," he wrote, "and so I said to myself 'Now, more than ever, must I continue with the search.'"[367]

When he synthesized and tested the DDT molecule, he observed, astonished, that the compound "showed a strong insecticidal contact action" to a degree he had never seen from any other substance. "My fly cage was so toxic after a short period that even after very thorough cleaning of the cage, untreated flies, on touching the walls, fell to the floor."[367] Müller determined that DDT satisfied all but one of his stringent criteria for the "ideal" insecticide; it fell short only on the criterion of "rapid onset of toxic action."

The first application of DDT was against the Colorado potato beetle in Switzerland in 1939; a patent on the compound followed in 1940.[366] Switzerland's neutrality during the war isolated the country, which boosted the importance of domestic food production and made especially welcome the increased potato yields due to DDT application.[363] In 1940, J. R. Geigy (Müller's employer) initiated sales of two products, a 5 percent DDT dust marketed as Gesarol spray insecticide for applications against potato beetles, and a 3 percent or 5 percent DDT dust marketed as Neocid dust insecticide for applications against lice.[63,368] The Swiss military was the first to use DDT for lice control on war refugees in order to prevent an outbreak of typhus.[363]

In the fall of 1942, J. R. Geigy relayed the discovery of DDT to both sides in the war. The Germans largely ignored DDT, though I. G. Farben (the company that produced Zyklon B for the termination camps, which was distributed by its subsidiary Degesch) manufactured DDT for use in louse powders.[303] The Nazis were more comfortable with their gas chamber technology and fixated on the industrial importance of Zyklon B.[333] But the US Army was interested, and it passed the samples of both compounds—Gesarol and Neocid—to scientists at the Department of Agriculture, who analyzed the chemical makeup, determined it was DDT, and resynthesized it.[349,358,364] (The British Ministry of Supply named the compound DDT in 1943, and the term stuck.)[63]

It then fell upon the US Bureau of Entomology and Plant Quarantine in Orlando, Florida, to study the efficacy of DDT and to confirm its toxic effect on flies.[349,369] The bureau had already tested seventy-five hundred chemicals for their ability to kill mosquitoes and lice.[370] But it took only a few days for the bureau to grasp the promise of this new chemical for disease control.[371] The bureau found that DDT killed flies, mosquitoes, lice, fleas, bedbugs, and many other insect pests by poisoning their nervous system, leading to the "Gesarol jitters" or the "DDTs." "On contact," wrote one enthusiastic journalist, "it causes paralysis of the limbs, spasmodic twitchings,

complete paralysis and then death."[372] In May 1943, DDT found its way onto the US Army supply lists.[373]

Critical experiments were then conducted to determine a safe mixture of DDT for application on human skin to kill body lice. It had already been determined via animal studies that high concentrations of DDT consumed with food caused serious health problems and even death, but it was not known if DDT could safely be applied to the skin.[358] Human tests proved highly successful. Brigadier General James Stevens Simmons wrote, "During field tests made with this powder among louse-infested natives in various parts of the world, it has been so popular that the investigators were frequently embarrassed by the large numbers of volunteers who demanded attention."[349]

Ideal doses were determined via a controlled study of thirty-five conscientious objectors to the war, carried out in New Hampshire's White Mountain Forest in the summer of 1943.[358] One hundred body lice were placed in the underpants of each individual and allowed to reproduce; soon, each participant hosted thousands of cooties, which were carefully counted daily. The participants, that "quiet group of college professors, farmers, clerks, salesmen, artists and professional men—conscientious objectors all—who had voluntarily journeyed to that camp in the remote section of the White Mountains for that express experience," were not permitted to scratch themselves.[358]

Every two weeks, each volunteer received a dose of DDT in a variety of carefully controlled mixtures of bases. Researchers then monitored the results. The lice descended first into "a sort of nervous agitation, then into a positive 'drunk,' then into a state of paralysis and finally into a coma which ended in death."[358] In customary government practice, the volunteers were never told the results of the experiment. However, the DDT dosing guidelines for skin application developed that summer in New Hampshire proved critically important just a few months later in the battle against typhus. (Such experiments on conscientious objectors were also invoked by

defense counsels in the Nuremberg war crimes tribunals.)[287] Once it was determined that DDT was safe for humans when used properly, the Headquarters Army Service Forces, the surgeon general, the quartermaster general, and the War Production Board put into effect a large-scale production effort.[349]

DDT experiments fell under a blanket of military secrecy. As long as the enemy did not have knowledge of DDT's effectiveness, enemy troops would continue to suffer casualties from malaria, yellow fever, typhus, and other insect-borne diseases. But the widespread use of DDT in the field inevitably leaked to the media. In a 1944 year-end review of science entitled "The Year Saw Many Discoveries and Advances Hastened by the Demands of the War," a journalist wrote, "What goes on in government and industrial laboratories and on the front these days is a military secret. Sometimes news trickles out that cannot be suppressed. It was thus that we learned of the advances made in jet-propulsion, winged rocket bombs, airplanes and DDT."[374]

The news about DDT appeared first in a trade journal in July 1943 with the publication of the structure of the molecule and the procedures for its synthesis, but this did not attract public attention.[303] That attention came on February 22, 1944, when the *New York Times* reported that the War Department had halted a typhus outbreak in Naples using a "new delousing powder known as DDT [which] was recognized as the greatest single weapon in the war against the disease."[375] The outbreak began in the overcrowded bomb shelters "where rats and vermin swarmed," conditions were filthy, and lice ubiquitous.[364] "Typhus," wrote Brigadier General Simmons, "prefers cold and temperate climates, and it burrows about in its filthy endemic lairs until disaster affords a chance to attack; then, through its loathsome intermediary, the louse, it preys on the miserable and the weak."[349]

The chief of the US Army Section of Preventive Medicine for the North African Theater of Operations, Colonel William S. Stone, mobilized his resources, and with the help of the Rockefeller Foundation health team and other medical officers, he achieved what had

never been done before: halting a typhus epidemic.[349] The goal was not simply to avert a humanitarian crisis; it was also to protect the American Fifth Army and Allied British, French, Canadian, and Polish soldiers.[358] In two months following December 26, 1943, the team blew DDT powder on nearly 2 million people in Naples at 40 delousing stations at a rate of 50,000 people per day.[364,376,377] They also created isolation hospitals, vaccinated people exposed to the sick, and sprayed DDT throughout the extensive network of six hundred caverns in which twenty thousand Neapolitans had hid from the bombing of their city for nine months.[358] "Neapolitans are now throwing DDT at brides instead of rice," reported the *New York Times*.[376] "Maybe it is because nobody in Italy is wasting food these days; maybe it is because of gratitude."

In all, only 1,377 cases of typhus occurred in Naples between December 1943 and February 1944, and no American soldiers died of the disease, largely because of the DDT control program coming on the heels of application of older insecticides that were available first.[320,358] Observers found these results "breathtaking" compared with the typhus outbreak that had occurred during World War I, less than three decades earlier, in which 9 million people died in Ukraine and the Balkans.

Delousing stations then cropped up throughout the expanding territories under Allied control. By April 1945, the Allies had created a "sanitary blockade" along the Rhine to halt the spread of typhus from Germany. "German civilians, displaced persons and released prisoners," reported the *New York Times*, "are not allowed to cross the rivers without first being examined and dusted with DDT powder."[378] A single application of DDT to undergarments of lice-infested people kept them free of lice for a month.[377] This monumental effort protected civilians and soldiers. "Typhus," wrote one journalist, "more dreaded than bullets in any army, is now simply unknown among our soldiers and sailors."[376]

In October 1945, the British evacuated fifty thousand German children from Berlin, "the over-crowded, desolated city, which lives from hand to mouth day to day."[379] Under Operation Stork, the

British evacuated two thousand children per day to the countryside to minimize the death rate during the approaching winter; each child was dusted with DDT powder to prevent typhus.

The wartime use of DDT was not limited to the battle against typhus. The efficacy of DDT as a residual pesticide against adult mosquitoes was also demonstrated in August 1943.[380] Researchers sprayed the inside surfaces of buildings and found that DDT killed *Anopheles*. The first field test against malaria occurred in May 1944 in Castel Volturno, a town north of Naples in Italy.[63] The Malaria Control Demonstration Unit of the Malaria Control Branch of the Public Health Sub-Commission of the Allied Control Commission sprayed the interior surfaces of all houses and structures in the town to test the effects on *Anopheles* and the incidence of malaria.[381] The next experiment was initiated in the Tiber delta, and both studies continued for two years. These studies were conducted by personnel of the same Rockefeller Foundation Health Commission that had successfully eradicated typhus in Naples a few months earlier. Many other experiments quickly followed.

By the end of July 1944, the US Army surgeon general, Major General Norman T. Kirk, announced that DDT, "one of the greatest discoveries of modern times," would be used to eradicate malaria-carrying mosquitoes.[361,382] After observing the aerial application of DDT onto the German-created malaria swamps in Italy, a reporter wrote, "It was an exciting experience to stand on the coast of the ancient Latium, chief base of Roman seapower in the Punic Wars, and observe the results of the experiment with this wonder-working chemical."[361] These first large-scale experiments accomplished, the army deployed DDT on an enormous scale in all war theaters. "Army preventive medicine," wrote Brigadier General Simmons, "is also smashing ahead on all fronts in the stubbornly resisted fight against malaria."[349]

It was thus that in 1944, through a combination of prevention and control measures, the infection rate of malaria in American troops fell to between one-quarter and one-third what it had been at the outset of the war.[383] In New Guinea, the malaria infection

rate for Allied soldiers fell from 3,300 cases per 1,000 troops per year in January 1943, to 31 cases per 1,000 troops per year in January 1944.[357] By the end of the war, the annual death rate of American soldiers for all diseases was 0.6 percent, far better than the 15.6 percent in World War I, and "lower than that ever attained by any armed force in the history of warfare."[384] This remarkable achievement was due to penicillin, atabrine, new anesthetics like metycaine, fibrin foam, concentrated plasma, new surgical techniques, and DDT.

Prime Minister Winston Churchill, in a speech to the House of Commons on September 28, 1944, bemoaned the 237,000 sick soldiers of the British Imperial Army fighting the Japanese in Burma.[385] But he found solace in the "excellent DDT powder, which has been fully experimented with and has been found to yield astonishing results." It will, he said, "henceforth be used on a great scale by the British forces in Burma and by the American and Australian forces, in India and in all theatres." Churchill pointed out that the Japanese also suffered "jungle diseases" and malaria, "which is an offset against the very heavy losses entailed upon our Indian, white and African troops." Churchill assured the House of Commons that "the war against the Japanese—and other diseases of the jungle—will be pressed forward with the utmost energy."

Just a few months after the successful malaria eradication in Italy, in December 1944, the army employed torpedo bombers, flying at 125 miles per hour at an altitude of 150 feet, to spray a 6,400 acre Pacific island with DDT at 2 quarts of DDT solution per acre. The goal was to prevent malaria among Allied troops, who had just conquered the island. "DDT also has been used in great quantities on the island's more than 7,000 Japanese corpses counted and buried thus far," wrote a correspondent.[386]

Such operations were possible not only because of the advent of DDT, but also because equipment developed by the US Chemical Warfare Service for poison gas dispersal, and already deployed in the war theaters, was readily adapted for DDT and other insecticide applications.[303] For example, the M-10 smoke tank only required a

new nozzle to spray DDT, and decontamination sprayers used to neutralize poison gas residues also successfully sprayed DDT.

One complication was that, from a medical point of view, spraying DDT onto beachheads before landing troops for combat would have been prudent, but the enemy might misinterpret the pesticide application as chemical or biological warfare.[303] This might have led to first use of chemical or biological weapons by the enemy, though, in their minds, such use would have been a retaliatory strike. A spiraling tit-for-tat conflict would have followed, as it had during World War I. Hence, DDT applications by Allied forces typically occurred following troop invasions.

Brigadier General Raymond W. Bliss, the assistant surgeon general, said upon his visit to Saipan Island, where American soldiers had suffered an epidemic of dengue fever, that he was not surprised that "about 8,000 Japs had been killed," but he was astonished by the complete absence of mosquitoes and flies.[387] When American forces first conquered Pacific islands, he said, clouds of insects made it difficult to see. "Now if one mosquito is located, we consider it comparable to finding a four-leaf clover." "The ghosts of the first Marines," wrote a journalist, "if they linger at Guadalcanal today, must smile with tolerant amusement at the changes two years have wrought."[388] The assistant commanding officer of the naval base told the journalist, "Hell, we don't even take atabrine anymore."

Americans greeted this parade of astonishing news, coming on the heels of heavy losses to tropical diseases early in the war, like soldiers coming home. "The reports of progress which reach the Surgeon General's Office almost daily," wrote Brigadier General Simmons, "compete in interest with the war bulletins from the fighting fronts . . . Such reports have fired the popular imagination, and the symbol DDT is acquiring a mysterious, romantic aura. It is coming so rapidly into common use that it bids fair to join the ranks of such well-known war-born Army terms as 'jeep,' 'radar,' and 'bazooka.'"[349]

In January 1944, US production of DDT amounted to less than 60,000 pounds per month; by the end of the year, it had skyrocketed to 2 million pounds per month, still almost exclusively for mili-

tary use, and still considered to be in critically short supply.[389,390] At the end of the war, US production reached 3 million pounds per month.[373] To provide the typhus control program in Naples in January 1944, DuPont charged $1.60 per pound.[391] By January 1945, due to the big production increase, the cost fell to 60 cents per pound. That month, Brigadier General Simmons wrote, "The possibilities of DDT are sufficient to stir the most sluggish imagination, but even if all investigations should cease today, we already have a proud record of achievement. In my opinion it is the war's greatest contribution to the future health of the world."[349]

Wartime advances against insect vectors of disease led to improvements beyond the development and widespread use of DDT and atabrine. Insect repellents were improved with the inclusion of dimethyl phthalate (DEET).[63] The airplane application of Paris green for larvicidal use was increased from a maximum payload of 700 pounds to one of 3,000 pounds, which greatly improved malaria control.[392] One particularly productive pilot dispersed more than half a million pounds of Paris green onto the island of Corsica in 1944. Pressure cylinders were created that used liquid Freon-12 as a propellant to disperse pyrethrum in a fine mist in order to kill mosquitoes in enclosed spaces—and 35 million of these bug bombs were deployed to the battlefield.[357]

Perhaps most remarkable, organizations with a history of on-again, off-again cooperation coordinated their efforts to eradicate vector-borne diseases, especially malaria. The US Army, US Navy, Public Health Service, National Research Council, Bureau of Entomology and Plant Quarantine, War Production Board, and Institute of Inter-American Affairs worked with each other and with chemical manufacturers, universities, and foundations to eradicate malaria on the battlefields. Cooperation also extended internationally among the United States, United Kingdom, and Australia.

These cooperative efforts and battlefield demands facilitated the scaling-up of DDT production to astronomical levels. By August 1945, Westinghouse alone produced 1.3 million DDT bug bombs per month for use in the Pacific Theater.[393] The bomb's lightweight

steel encasement contained 2 percent pyrethrum, 3 percent DDT, 5 percent cyclohexanone, 5 percent lubricating oil, and 85 percent Freon-12, which acted as the dispensing agent.[373] The bomb was innovative not just for the use of DDT, but also for this use of Freon to distribute the insecticide. One of the most prominent physicians in the United States wrote in 1945 that "the Freon bomb alone may, in the years to come, more than repay the costs of World War II."[394] Freon was later banned by the Montreal Protocol because it destroyed ozone in the Earth's stratosphere.

Meanwhile, the need for DDT to eradicate malaria in Europe was still urgent. Greece experienced a malaria epidemic immediately after the war ended there, with reports that in some areas all residents were infected.[395] In response, in August 1945, the UN Relief and Rehabilitation Administration initiated the largest-ever airborne combat against malaria.

Paul Müller and his colleague Paul Lauger, the director of research at J. R. Geigy, announced that same month, in their first public statement to the press, that DDT could prevent between 1 million and 3 million deaths from malaria per year, that "it might even rid the earth eventually of all insect-transmitted diseases," and that it could "ultimately eliminate all flies and mosquitos from the United States."[396] The *Anopheles* mosquito and the typhus-carrying louse would "join the dodo and the dinosaur in the limbo of extinct species."[303]

The Miracle of DDT (1945–50)

If the tales of DDT miracles are borne out, there ought to be no excuse for any insect-borne disease. A housefly ought to become a curiosity; dogs ought to lead flealess lives of bliss.—*New York Times* **report on the conquest of typhus in Naples, 1944**[376]

During the war, small experimental trials of DDT were allowed for civilian use. For example, on July 8, 1945, the Long Island State Park Commission and the Nassau County Mosquito Extermination

Commission tested the Todd Insecticidal Fog Applicator, an adaptation of the fog generator used on the war fronts, at Jones Beach, New York.[397] "Early arrivals among the 60,000 visitors at Jones Beach State Park here today found themselves suddenly enveloped in billowing clouds of sweetish smelling fumes," wrote a correspondent for the New York Times. "Mounted on an open truck, the device blanketed the beach at the rate of an acre a minute with a fog mixture of light oil, atomized into infinitesimal particles by the generator, and a 5 per cent solution of DDT." The fumes, noted the correspondent, "do not cause harm or discomfort to human beings." In fact, the model Kay Heffernon was employed to eat a hot dog and drink a Coke while in the midst of the DDT fog. The New York Times reported that "after the cloud had dissipated—it lasted only fifteen minutes—not a fly or mosquito remained."[372] The experiment demonstrated that at a cost of only 17 cents per acre, mosquitoes and biting flies could be eradicated from beaches.[398]

Near the end of the war, a close observer of DDT's series of successes on the battlefield wrote that "the post-war potentialities of DDT are almost infinite."[358] At the beginning of August 1945, the War Production Board allowed small quantities of DDT to be used for civilian and agricultural applications.[399] A few days later, the United States dropped a nuclear bomb on Hiroshima, and then another on Nagasaki. Time magazine coupled photos of the first atomic explosion on the same page with the news that DDT was available for civilians.[400]

Soon thereafter, with the war over and production no longer restricted for military use, American chemical companies expanded their output of DDT and directed it toward civilian life. In 1945, US production reached 36 million pounds.[102] That year, an advertisement for the insecticidal Flitgun, which had begun to incorporate DDT into its formulation, paired an image of a soldier shooting the Japanese enemy in the back with an image of a person spraying a fly with a Flitgun. The ad read, "Whether Japs or flies, it's fast action that counts."[303] By the late 1950s, annual US production equaled 1 pound per person, or 180 million pounds per year.[102]

Figs. 8.4a & 8.4b. Jones Beach DDT experiment, and demonstration with model Kay Heffernon, July 8, 1945. Photo of beach spraying from the Bettmann Collection; photo of Kay Heffernon by George Silk, from the LIFE Picture Collection; both via Getty Images.

Production on such a scale was possible in part because of the use of chemical warfare production facilities by insecticide manufacturers.[303] These facilities were purchased or leased from the Chemical Warfare Service, which Congress elevated into the Chemical Corps in 1946. A spectacular example of conversion of a war gas production facility to a private sector pesticide plant is the Rocky Mountain Arsenal; the plant for producing mustard gas was used by a new insecticide company to develop two new pesticides similar in structure to DDT: aldrin and dieldrin. Leadership of the new company was provided by the former commander.

In the United States, DDT and other new pesticides facilitated the rapid expansion of the chemical industry. In 1939, 83 US companies focused their manufacturing on insecticides and fungicides.[363] By 1954, the number had risen to 275.

The postwar flurry of excitement over DDT drew it into every corner of American life. DDT was even impregnated into wallpaper and paint, where it would "chase flies, mosquitos and other insects from a room."[401] DDT paint itself had many uses, from covering doorways, screens, garbage cans, and drains in houses to protecting the hulls of ships from barnacles.[402] Sherwin-Williams demonstrated the efficacy of DDT paint to one hundred fifty thousand people in a Cleveland exposition by releasing ten thousand flies and letting them alight on screens that had been painted weeks earlier; one by one the flies got the "DDTs" and died.[303]

In July 1945, the Connecticut Agricultural Experimental Station and the USDA Bureau of Entomology and Plant Quarantine sprayed the Yale Bowl via helicopter before a pop concert "so that music lovers could enjoy the program without mosquito-slapping."[403] The New Jersey Mosquito Extermination Association advocated, on March 30, 1945, the firing of mortars loaded with DDT over the swamps and meadows of the state.[404] An Army plane sprayed the Jersey salt marshes on August 4, 1945. On August 9, 1945, the Michigan state health department sprayed DDT over Mackinac Island.[405] A journalist noted that "extinction of the fly was celebrated here today. In a public bonfire hundreds of old flytraps were burned. Coachmen

Fig. 8.5. Companies incorporated DDT into a wide variety of consumer products, including wallpaper for nurseries.

who drive the island's famous surreys stowed away their horse nets . . . [DDT's] effect on flies was like the atomic bomb."

In mid-August 1945, a Mitchell bomber flying at 200 miles per hour at 150 feet sprayed 1,100 gallons of DDT over half of Rockford, Illinois, to halt the spread of infantile paralysis due to polio.[406,407] The purpose was to eradicate flies that might transfer the polio virus from human feces to food. The other half of the city was left untouched to allow the effectiveness of DDT to be evaluated, but only the authorities and observant residents knew which half of the city was the test area and which half the control. "To make public this information," said a polio expert from Yale University in

charge of the campaign, "might cause a stampede into the protected area. This and the hysteria that could attend such a movement are the things we want to prevent." This experiment inspired many others on the East Coast of the United States, led by the Fly Abatement Unit of the Army Neurotropic Virus Commission.[408]

Upscale department stores added DDT products to their lines of luxury goods. Macy's began selling MY-T-KIL insecticide in August 1945 for 49 cents a quart.[409] Bloomingdale's advertised a half-pint for $1.25.[410] By September, supply restrictions had eased sufficiently for the comforts afforded by DDT to be shared with zoo animals. The same insecticidal fog applicator used at Jones Beach sprayed a mixture of 3.5 pounds of DDT in 11 gallons of grade 2 fuel oil, with pine oil for scent, onto Chang the elephant, as well

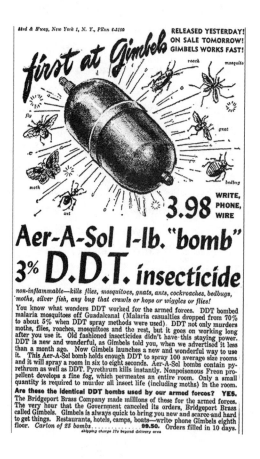

Fig. 8.6. The new and wonderful way of using DDT presented by the retailer Gimbels.

as the giraffes, bison, elk, and red deer at the Central Park Zoo.[411] The Department of Parks also pumped the fog into bird houses through the zoo's network of steam and water pipes. In this way, the flies and other bothersome insects were exterminated from the zoo.

Also in September, Westinghouse Electric Appliance Division announced that its DDT bug bomb used during the war would "become increasingly available in the near future for housewives who want to kill flies, mosquitos and other insects around the house."[412] Each bug bomb contained 1 pound of aerosol, enough to "debug" 150,000 cubic feet of space, the equivalent of 10–15 homes. Once the valve was opened, "the insecticide is propelled into the air in particles so fine that there may be 100,000,000 of them in a single drop of liquid." Conveniently, "housewives using the bomb do not have to wear any special protective clothing or mask."

With the explosion of DDT into the civilian marketplace, all manner of fraudulent claims were made. Barely a month after DDT products hit the shelves, in September 1945, the US Department of Agriculture launched a countrywide campaign against firms and individuals who violated the labeling provision of the Insecticide Act of 1910 by selling "so-called DDT products," which really contained as little as 0.01 percent DDT, rather than the recommended 5 percent.[413] The Better Business Bureau of New York City, responding to complaints about false advertisements, found two areas of concern: "the overemphasis of DDT in labels and copy, suggesting that it is the principal ingredient in an insecticide; and the claim, expressed or implied, that DDT will kill all insect pests."[414]

The availability of DDT constituted part of the overall easing of wartime restrictions. The Depression and the war bottled up American consumerism in a decades-long funk, and this pent-up enthusiasm to spend could not be held back any longer. A *New York Times* article recommended the "hand-grenade shaped" DDT aerosol bomb for a Christmas gift.[415] Macy's ran an advertisement on October 1, 1945, under the banner "When Will Macy*s Have It?"[416] Included in that list:

When are you going to have really good American watches?

We had Elgins and Hamiltons last week. More in a month or so.

How about lawnmowers?

We were the first with them weeks ago, and we have them now.

Men's white shirts with French cuffs?

Where were you last Thursday? Keep watching for more.

Airplanes?

Within a matter of days.

Vacuum cleaners?

Samples in a few weeks. Good stocks probably by November.

DDT Aerosol "bombs"?

New low price at Macy's.

Phonographs?

On Aug. 6th we got in more than we'd had in 4 years. Still a wide choice today.

Sheets?

We have 'em—we just don't talk about 'em. In 3 or 4 weeks, we'll talk!

Washing machines?

If not October—then certainly November. And *not* a preview: you'll be able to *buy!*

Harmonicas?

They're back! We were first to have metal harmonicas on Sept. 14.

DDT's power bestowed upon it metaphorical status. New York mayor Fiorello H. La Guardia complimented the 1945 No Deal Party candidate for mayor, Newbold Morris, upon his oratory skills by describing his radio talk as "a most impressive, telling statement, fortified by sincerity, it was indeed a skillful use of heavy artillery and a beneficial sprinkling of DDT at one and the same time."[417] DDT even crept into news stories about the prosecution of war criminals. "Omori prison camp, that Nippon pest hole," wrote a correspondent, "has been cleaned up with DDT powder, and General Tojo and other high-toned Jap war criminals have moved in. To deverminate it now will call for a stronger kind of powder."[418]

Fig. 8.7. Theodor S. Geisel (Dr. Seuss) illustrated advertisements for the insecticidal Flit gun both before the war and afterward, when DDT was incorporated into the product.

Production increases led to further price decreases. In 1946, just one year after the war, the US federal government sprayed DDT over vast tracts of land to eradicate the gypsy moth at a cost of only $1.45 per acre, whereas the cost of dispensing the much less effective arsenate of lead from high-powered ground sprayers reached up to $25 per acre.[419] The amateur entomologist Leopold Trouvelot had brought gypsy moths from France to Massachusetts in the late 1860s for use in his hobby of crossbreeding silk-making caterpillars.[320] Larvae escaped from Trouvelot's backyard and spread first down his street (coinciding with his move back to France) and then throughout much of the eastern United States, destroying valuable trees in their path.

By 1946, twenty-five American companies were already in the business of making DDT aerosol bombs.[303] DDT composed part of the wave of the postwar, post-Depression economic boom. It stimulated a flurry of jobs in production, distribution, marketing, sales, application, and health care.

The list of arthropod pests against which DDT was effective looked like an entomologist's record of lifetime collections. The prominent entomologist Clay Lyle, writing to professionals in the

field in 1947, described the status of DDT this way: "Suffice it to say that at no previous time in history have the achievements of entomologists, working in collaboration with chemists and engineers, been of such universal value as to make in so short a time the name of an insecticide a common word in every household however humble or remote. The entomologist has become a wizard in the eyes of the uninitiated—and indeed some of the achievements seem little short of magic . . . Is not this an auspicious time for entomologists to launch determined campaigns for the complete extermination of some of the pests which have plagued man through the ages? . . . Let us not be satisfied with anything less than a post-war program which will challenge the imagination of the world."[419]

DDT became an essential agricultural commodity immediately after the war. In Washington, apple growers swapped their use of lead arsenate and sodium aluminofluoride (cryolite) for DDT and were rewarded with a huge decline in losses from coddling moths.[363] In Kansas, ranchers calculated that each pound of DDT used for fly control resulted in an extra 2,000 pounds of meat production. Such reports of amazing agricultural yields due to DDT seemed to come from all manner of food production. In 1920, each American farm worker grew enough food for eight people, but by 1957 that number reached twenty-three.[420] Consumers also benefited from DDT as a replacement pesticide because the residues of lead arsenate and other metal-based pesticides on unwashed produce had caused numerous poisonings year after year.[421]

Idaho initiated a fly eradication program using DDT whose slogan, "No Flies in Idaho," headlined posters in ten thousand public places.[419] "If the campaigns are pushed vigorously early next spring," predicted Lyle, "there should be many situations before the end of 1947 where a city or county health officer will be asking anybody who has seen a fly to report it, after which a spraying crew will visit the property to locate and destroy breeding sources. This is not a fantastic dream but is something that is almost certain to happen."[419]

Similar programs sprang up around the United States, such that by 1950 six hundred cities in forty-five states employed fog

generators developed by the Chemical Warfare Service to disperse DDT.[303] A glut of military aircraft and veteran pilots combined with cheap DDT to revitalize the crop dusting business. DDT had penetrated municipal, state, and federal insect control programs, agriculture, and the home.

Shortly after the war, other synthetic insecticides also entered the marketplace, such as the chlorinated compounds chlordane, toxaphene, lindane, BHC, and methoxychlor, and the organophosphate compound parathion—a total of twenty-five new pesticides by 1953.[363] Chlordane was hailed as "the greatest discovery since DDT."[303] Sales were robust, and pesticide plants operated around the clock to meet demand. "In the beginning all these insecticides were greeted with great optimism," wrote Müller, "and people prophesied only a short life for dichloro-diphenyl-trichloroethane [DDT]. On the whole there is now less talk of [these new insecticides] and DDT has, particularly in the field of hygiene, maintained and even improved its dominating position."[367]

Indeed, DDT continued its march against malaria throughout the world. In the United States in July 1944, Brigadier General Simmons successfully advocated for an aggressive domestic campaign against malaria through an expansion of the Malaria Control in War Areas program.[303] The program morphed in 1946 into the Communicable Disease Center, later to become the Centers for Disease Control and Prevention in Atlanta.

The British employed DDT in antimalaria campaigns in Britain's African colonies, where they had to overcome resistance by suspicious elders who questioned its safety. In the 1946 antimalaria campaign in the Kipsigi Tribal Reserve in Kenya, Kipsigi leaders greeted with skepticism assurances by the British entomologist in charge of demonstrating efficacy and safety. "The Africans at first are not very impressed," stated the narrator of the documentary about this effort.[422] "Some are afraid that the DDT will poison them, while others suspect some form of witchcraft." In response, the entomologist ate a bowl of porridge sprayed with DDT, but "even this fails to convince the audience." One elder stated that it was a bad poison that

Fig. 8.8. "DDT is a benefactor of all humanity." This Pennsylvania Salt Manufacturing Company advertisement appeared in the June 30, 1947, issue of *Time* magazine.

might exterminate the whole tribe. Yet DDT gained acceptance in Kenya as it did elsewhere; the lethal effects of this new insecticide on insect pests were undeniable, and the door was thrown open to simultaneously conquering disease and alleviating hunger.

DDT stacked up remarkable public health achievements in country after country. It was employed, along with the rodenticide sodium fluoroacetate ("1080"), to immediately halt a bubonic plague

outbreak in Peru from December 1945 to January 1946.[423] It was credited for a large measure of the 50 percent decline in the death rate in India in the two decades after its release, and in Ceylon (Sri Lanka) the program of spraying households with DDT was followed by a 34 percent decline in deaths in only one year.[424]

Throughout the world, people embraced the opportunities afforded by DDT in the realms of public health, food security, and pest management in the home. A panacea that addressed problems ranging from plagues to household annoyances, and at such a low cost, struck ministers of health and housewives alike as exemplifying the limitlessness of human ingenuity. Within twenty-three years from the end of the war, enthusiastic chemical companies and consumers had produced and sprayed 1 billion pounds of DDT.[425] Never before in human history had a new chemical been so widely broadcast.[421] The success of DDT initiated a flurry of activity in chemistry labs around the world, and scientists developed an immense diversity of synthetic pesticides in rapid succession.[370] These pesticides fueled the Green Revolution, corporate profits, and the reputations of scientists like Müller credited with preventing vector-borne diseases and famine.

[**9**]

I. G. Farben
(1916–1959)

Behold, this is the enemy of the world, the destroyer of civilizations, the parasite among nations, the son of chaos, evil incarnate, the ferment of decay, the formative demon of humanity's destruction.—**Reich Minister of Propaganda Joseph Goebbels at the Nuremberg Nazi Party rally, September 9, 1937**[305]

God gave us this earth to be cultivated as a garden, not to be turned into a stinking pile of rubble and refuse.—**Brigadier General Telford Taylor,** chief prosecutor in the US Military Tribunal at Nuremberg against I. G. Farben officials, August 27, 1947[426]

After World War I, many critics accused chemical companies not only of profiteering from war, but also of encouraging it.[303] In 1934, *Fortune* magazine published a commentary reflecting on the cost of killing each enemy combatant in the war, which it calculated at $25,000.[427] *Fortune* stated that "every time a burst shell fragment finds its way into the brain, the heart, or the intestines of a man in the front line, a great part of the $25,000, much of it profit, finds its way into the pocket of the armament maker."[427] A French economist pointed out that the role of competition in the armaments

industry is unique: "The trade in arms is the only one in which an order obtained by a competitor increases that of his rivals. The great armament firms of hostile powers oppose one another like pillars supporting the same arch. And the opposition of their governments makes their common prosperity."[427]

Senator Gerald Nye of North Dakota republished the *Fortune* article in the *Congressional Record*, and it was condensed for the subscribers of *Reader's Digest*.[303] The bestselling book *Merchants of Death*, which emphasized that World War I created 21,000 new millionaires in the United States, absorbed readers that same year, and many other, similar publications followed.[428] Senator Nye chaired a Senate committee to investigate munitions companies, prompting a member of the Du Pont family to accuse communists of inciting the popular reaction against his and other chemical companies. The DuPont company had garnered profits of $228 million during World War I, and so the company took the brunt of the criticism.[303] Nye quipped, "We may expect the next war to make the world safe for du Pontcracy."[429]

President Franklin D. Roosevelt expressed a similar sentiment when he stated, "The private and uncontrolled manufacture of arms and munitions and the traffic therein have become a serious source of international discord and strife . . . This grave menace to the peace of the world is due in no small measure to the uncontrolled activities of the manufacturers and merchants of engines of destruction, and it must be met by the concerted action of the peoples of all nations."[430]

Advances in chemical weaponry and insecticides continued to feed into each other's development in the years leading up to World War II. This was especially the case in Germany. The German chemical conglomerate I. G. was formed in 1916 as a "community of interest of the German dyestuff industry" (Interessen Gemeinschaft der Deutschen Teerfarbenindustrie) from the chemical giants BASF, Bayer, and Hoechst, along with the smaller companies Agfa, Cassella, Kalle, Ter Meer, and Greisham.[326]

In 1925 these eight companies combined their postwar resources into a single corporation known as I. G. Farbenindustrie Aktiengesellschaft, or I. G. Farben for short.[326] It was the largest chemical company in the world, and its share value more than tripled the following year. The company expanded into munitions and so integrated its nitrate plants with the consumers of those nitrates in the explosives industry. It also pushed an aggressive production program for synthetic oil from coal, despite the high cost, because Germany's lack of domestic oil production posed a major war liability.

With the ascent of the Nazis, I. G. Farben cozied up to the party leadership, and provided critical financial support to Hitler's election campaign in early 1933.[326] I. G. Farben was Germany's largest corporation, and the company provided Hitler with the single largest donation in his election campaign.

At first the company leadership aimed to protect its Jewish scientists.[326] Carl Bosch, the first engineer to win a Nobel Prize in Chemistry for his work on chemical high pressure methods, met with Hitler just after the March 1933 election in order to discuss the importance of I. G. Farben's program to synthesize oil from coal. He told Hitler that the expulsion of Jewish scientists from Germany would set back both physics and chemistry in the country by a century. Hitler immediately responded, "Then we'll work a hundred years without physics and chemistry!"[326] Hitler never again met with Bosch.

Bosch learned in April 1933 that his former partner Haber had been forced to resign his professorship at Berlin University and his directorship of the Kaiser Wilhelm Institute for Physical Chemistry and Electrochemistry, even though he was a convert to Christianity and one of the most prominent German scientists.[326] Bosch tried to gather Germany's non-Jewish Nobel Prize winners into a group that would resist the persecution of Jewish scientists. The effort failed as even one of Haber's students said, "We cannot draw our swords for the Jews!"[301]

I. G. Farben completed its Nazification in 1937.[326] Nearly all board members who were not yet in the party joined; Jewish officials, including supervisory board members, were fired; and Bosch fell from his post as leader of the company to an honorary position. The company led corporate financial backing of the Nazi Party, and it provided the war materials that enabled Nazi expansionism: synthetic oil, synthetic rubber (buna), lubricants, explosives, plasticizers, dyestuffs, and thousands of other chemically based essential ingredients of war, including poison gases.

The fruits of that expansion for I. G. Farben began with the annexation of Austria on March 11, 1938.[326] Within days the company sent a memo to Nazi occupation officials in which I. G. Farben asked for the authority to absorb Austria's largest chemical company, Pulverfabrik Skodawerke-Wetzler A.G. (Skoda-Wetzler Works).[326,426] This takeover, I. G. Farben officials argued, would strike a blow to Jewish influence in Austrian industry because the Rothschild family held the majority share interest in the bank that controlled the company. By the fall of 1938, Jewish leaders of Skoda had been fired, the general manager of Skoda had been stomped to death by Nazi storm troopers, the Jewish representative of the Rothschilds had fled Austria, and the company belonged to I. G. Farben.[326]

Next to fall to Nazi and I. G. Farben expansionism was Czechoslovakia, with the September 29, 1938, signing of the Munich Pact by the British prime minister Neville Chamberlain and the French prime minister Edouard Daladier.[326] This capitulation to Nazi demands for control of Sudetenland, the German-speaking regions of Czechoslovakia, also facilitated I. G. Farben's scheme to absorb the largest chemical company in Czechoslovakia, the Verein für Chemische und Metallurgische Produktion of Prague (Prager Verein).[426]

Because 25 percent of the Czech company's directors were Jewish, I. G. Farben had the leverage it needed for the takeover.[326] The Sudeten-German Economic Board confirmed this political reality with its advisory to I. G. Farben that the "Czech-Jewish management in Prague is done for."[426] The day after the signing of the Munich Pact, the head of I. G. Farben, Hermann Schmitz, sent a telegram to

Hitler hinting at the company's interest in the region. Schmitz was, he wrote, "profoundly impressed by the return of Sudeten-Germany to the Reich, which you, my Führer, have achieved."[426] Schmitz added to the telegram that I. G. Farben "puts an amount of half a million reichsmarks at your disposal for use in the Sudeten-German territory."[426]

German troops entered Sudetenland the following day, on October 1, 1938.[426] "Negotiations" for the takeover of Prager Verein stalled as company management resisted. At a meeting on December 8, an I. G. Farben leader, Georg von Schnitzler, told the Prager Verein leadership that he knew they were undermining the deal for I. G. Farben to acquire their critical chemical plants. He would, therefore, need to alert the German government that because of their attitude, "social peace in the Sudeten area was being menaced and that unrest could be expected at any moment."[426] The responsibility for that unrest, he informed them, would be theirs. The following day, the Prager Verein leadership agreed to sell the critical plants.

Poland was Germany's next target. As with the prior conquests, in advance of the invasion, I. G. Farben prepared a wish list of chemical companies it would like to acquire in Poland.[326,426] Germany invaded on September 1, 1939, and so began World War II. Three weeks later, officials of I. G. Farben, at their request, were appointed trustees of the absorbed Polish chemical plants.[426]

Bosch fell into depression contemplating how German aggression and the war had resulted from his chemical engineering successes of synthesizing nitrates, oil, and rubber.[326] In February 1940 he took his ant colony from the Kaiser Wilhelm Institute and moved to Sicily. His health continued to decline, and he returned to Germany in April. He predicted the fall of France and the ultimate destruction of Germany and the company he had built. On April 26, 1940, at the age of sixty-five, Bosch died.

Germany invaded France on May 9, and by late June Germany controlled nearly all of Europe.[326] I. G. Farben absorbed the chemical companies of France, Norway, Holland, Denmark, Luxembourg,

and Belgium. It planned for the conquest of chemical companies in countries not yet conquered, whether neutral, ally, or enemy, including the Soviet Union, Switzerland, the United Kingdom, Italy, and the United States. "I. G. was at the zenith of its power," wrote an expert on the company. "From the Barents Sea to the Mediterranean, from the Channel Islands to Auschwitz, it exercised control over an industrial empire the likes of which the world had never before seen."[326]

If the leaders of chemical companies in conquered territories were unwilling to cede a controlling interest to I. G. Farben, they were threatened with classification as a "Jewish concern," which would allow immediate and complete confiscation of assets.[326] The French chemical giant Kuhlmann, which was Europe's second-largest chemical company, served an example. Kuhlmann had been managed by a Jew prior to the German occupation; this fact alone was sufficient for classification as a Jewish concern and subsequent seizure, which compelled the Kuhlmann leadership to capitulate to I. G. Farben's demands.

I. G. Farben was complicit in the Holocaust not only for the plunder of Jewish assets and for the production of Zyklon B for the termination camps, but also for the testing of chemicals on concentration camp inmates.[426] An I. G. Farben director later justified the human experiments on the grounds that "the inmates of concentration camps would have been killed anyway by the Nazis" and "the experiments had a humanitarian aspect in that the lives of countless German workers were saved thereby."[431]

Perhaps the greatest crime of I. G. Farben was its use of slave labor. The Nazis and their corporate allies utilized slaves from concentration camps throughout their sprawling network of production facilities for war materials and weapons. Slaves were cheap, expendable, and ensured secrecy. As Heinrich Himmler explained to Hitler in their planning for the use of slave labor, a benefit of using concentration camp slaves would be that "all contact with the outside world would be eliminated. Prisoners don't even receive mail."[362]

The I. G. Farben chemist Otto Ambros selected the Auschwitz concentration camp as the location for the company's extensive manufacturing complex to produce synthetic rubber and oil—the largest such facility in the world and I. G. Farben's largest endeavor.[326,426] Ambros selected Auschwitz because of the availability of slave labor, a coal mine, plenty of water (the Sola, Vistula, and Przemsza Rivers), and a rail and road network.

Ambros negotiated with Himmler for the "rent" of Auschwitz slave labor in the Farben facilities—negotiations facilitated by the mutual benefit of such an arrangement to I. G. Farben and the SS, and the fact that the two men had known each other since attending grade school together.[362] Farben agreed to pay the SS 3 Reichsmarks per day per slave. Ambros crooned about the arrangement to his boss, writing, "On the occasion of a dinner party given for us by the management of the concentration camp, we furthermore determined all the arrangements relating to the involvement of the really excellent concentration-camp operation in support of the Buna plants. Our new friendship with the SS is proving very profitable."[362] Ambros chose to participate in the Holocaust despite the fact that his PhD supervisor was Haber's friend, the Jewish Nobel Laureate Richard Willstätter, with whom Ambros corresponded even after Willstätter fled to Switzerland in 1939; Willstätter died in exile in 1942.[305,326]

The I. G. Farben complex at Auschwitz, designated as "I. G. Auschwitz," consumed more electricity than Berlin.[326] Twenty-five thousand concentration camp inmates died building the complex. Because inmates often died on their forced march from the main Auschwitz concentration camp to the construction site, and this reduced productivity, I. G. Farben built its own concentration camp for slave labor. This camp, called Monowitz, also bore the Auschwitz motto across its entrance: *Arbeit macht frei*, or "Work sets you free." The entire Auschwitz complex was then complete: Auschwitz I, the original concentration camp containing hundreds of thousands of prisoners; Auschwitz II, the extermination camp at Birkenau; Auschwitz III, the I. G. Farben facilities for the production of rubber and

fuel; and Auschwitz IV, the I. G. Farben concentration camp at Monowitz.

I. G. Farben officials were unhappy with the labor shortage caused by the "selection" process of SS physicians; Jews capable of work in the I. G. Farben facility were too often selected for gassing at Birkenau due to the zeal of the SS to carry out the Final Solution.[326] For example, in a shipment of 5,022 Jews to Auschwitz, 81 percent were selected for gassing, while only 19 percent were selected to labor for I. G. Farben. An SS official, seeking to increase the labor pool for I. G. Farben, altered the protocol such that the unloading of Jews from trains would occur near the I. G. facility rather than near the crematory. The next shipment of 4,087 Jews resulted in 59 percent selected for the gas chambers and 41 percent selected for I. G. Farben labor, thereby improving the labor statistics. Between 1941 and 1945, I. G. Farben enslaved 275,000 concentration camp inmates, excluding those who died or were exchanged with other slaveholding businesses.[426]

I. G. Farben officials were continually unsatisfied with the labor pool. "If the transports from Berlin continue to have so many women and children as well as old Jews," said an official, "I don't promise myself much in the matter of labor allocation."[326] For the typical Auschwitz inmate selected for I. G. Farben work, Monowitz proved to be the location for a few months of labor before their poor condition put them on the path to Birkenau.[326] An SS officer told the prisoners of Monowitz, "You are all condemned to die, but the execution of your sentence will take a little while."[326]

I. G. Farben prisoners at Auschwitz received better food than at the extermination camps, but it was still a starvation diet that resulted in a weight loss of 3–4 kg per week.[326] I. G. Farben officials imposed a rule that no more than 5 percent of its inmates could be sick at any given time and a single individual could not be sick for more than fourteen days. Officials managed any exceedance of these rules by shipping the sick to Birkenau.

I. G. Farben foremen ensured discipline among their slave laborers by referring rule breaches to the SS.[326] Examples of transgres-

sions that received SS discipline included "lazy," "shirking," "refusal
to obey," "slow to obey," "working too slowly," "eating bones from a
garbage pail," "begging bread from prisoners of war," "smoking a ciga-
rette," "leaving work for ten minutes," "sitting during working hours,"
"stealing wood for a fire," "stealing a kettle of soup," "possession of
money," "talking to a female inmate," and "warming hands."[326] The SS
responded with a range of punishments, including withholding food,
lashing, hanging, and selecting the offender for Birkenau.

On January 17, 1945, as Russia's Red Army advanced toward
Auschwitz, Ambros busied himself with the destruction of documents related to I. G. Farben's wartime activities and atrocities.[362]
The next day, SS guards forced the remaining Monowitz prisoners
on a death march into the German interior; within two days, 60 percent of the prisoners were dead. On January 23, Ambros departed
from Monowitz, leaving behind only those prisoners with infectious diseases who had been too weak to force on the death march.
One of those was Primo Levi, the Italian chemist and writer. Four
days later, on January 27, the Red Army liberated what was left of
the Auschwitz concentration camps. Meanwhile, Ambros traveled
back to Germany to destroy other I. G. Farben records and to mod-
ify a chemical weapons factory to appear as if its purpose was to
produce detergents and soap.

The US Military Tribunal at Nuremberg indicted twenty-four
I. G. Farben leaders for war crimes.[426] The most serious charge, "Count
Three—Slavery and Mass Murder," stated: "In the course of these
activities, millions of persons were uprooted from their homes, de-
ported, enslaved, ill-treated, terrorized, tortured, and murdered."[426]
Brigadier General Telford Taylor, the chief prosecutor, opened the
case on August 27, 1947, with a chilling statement to the court:

> The grave charges in this case have not been laid before
> the Tribunal casually or unreflectingly. The indictment accuses
> these men of major responsibility for visiting upon mankind
> the most searing and catastrophic war in human history. It
> accuses them of wholesale enslavement, plunder, and murder.

These are terrible charges; no man should underwrite them frivolously or vengefully, or without deep and humble awareness of the responsibility which he thereby shoulders. There is no laughter in this case; neither is there any hate.[426]

General Taylor stressed the methodical nature of the actions of I. G. Farben's leaders:

The crimes with which these men are charged were not committed in rage, or under the stress of sudden temptation; they were not the slips or lapses of otherwise well-ordered men. One does not build a stupendous war machine in a fit of passion, nor an Auschwitz factory during a passing spasm of brutality. What these men did was done with the utmost deliberation . . . In this arrogant and supremely criminal adventure, the defendants were eager and leading participants. They joined in stamping out the flame of liberty, and in subjecting the German people to the monstrous, grinding tyranny of the Third Reich, whose purpose it was to brutalize the nation and fill the people with hate. They marshaled their imperial resources and focused their formidable talents to forge the weapons and other implements of conquest which spread the German terror. They were the warp and woof of the dark mantle of death that settled over Europe.[426]

General Dwight D. Eisenhower, at the close of the war, assigned an investigative team to explore I. G. Farben's role in the creation of Nazi military power, including I. G. Farben's absorption of the chemical companies of conquered nations.[326] General Taylor noted in the Nuremberg trials that "Europe was dotted with mines and factories which they coveted, and for each step in the march of conquest there was a program of industrial plunder which was put into prompt and ruthless execution."[426] Eisenhower's team concluded, "Without I. G.'s immense productive facilities, its far-reaching re-

PART 3

War

search, varied technical experience and overall concentration of economic power, Germany would not have been in a position to start its aggressive war in September 1939."[326]

The court sentenced Ambros to eight years in prison for "war crimes and crimes against humanity through participation in enslavement and forced labor" as well as "the mistreatment, terrorization, torture and murder of enslaved persons."[432] Eleven other I. G. Farben officials were found guilty and sentenced to imprisonment for periods of one and one-half years to eight years.[326] An outraged chief prosecutor said the punishments were "light enough to please a chicken thief."[326]

Tabun (1936–45)

Mix together saltpetre, [charcoal], and sulphur, and you will make thunder and lightning.—**Roger Bacon**, ca. 1270s[433]

We wish, at the moment, to poison insects because they threaten the health of our troops. Coincidentally the Chemical Warfare Service . . . is actively at work in an attempt to improve our methods for poisoning Germans and Japanese . . . The fundamental biological principles of poisoning Japanese, insects, rats, bacteria, and cancer are essentially the same. Basic information developed concerning any one of these topics is certain to apply to the others.—**William N. Porter**, Chemical Warfare Service Chief, 1944[303]

With the ascent of the Nazis to total power in Germany and before the outbreak of war, I. G. Farben officials advocated enhanced German production of chemical weapons. These, they argued, could be deployed against enemy civilians in the upcoming war to instill panic when people would find "every door-handle, every fence, every paving stone a weapon."[400] I. G. Farben officials pointed out that Germans would hold the advantage even if Allied forces retaliated because of the storied German discipline.

The German motivation for chemical weapons production also grew out of the economic burden of importing expensive insecticides, primarily nicotine.[400,434] German law in 1937 mandated insecticide use by farmers, so German chemical companies could profit from sales to farmers if they could discover inexpensive chemicals that were toxic to insects. They could also profit from sales to the military if, in the course of their pesticide research, they discovered chemicals toxic to people.

The I. G. Farben chemist Gerhard Schrader critically advanced both fronts when, in his search for new insecticides, he manipulated the structure of the toxic compound chloroethyl alcohol.[435] Schrader replaced different atoms in the molecule and tested the resulting compounds for toxicity. He arrived at a family of chemicals known as organophosphates and found many to be extremely toxic to insects.[303,370,435]

On December 23, 1936, Schrader synthesized an organophosphate chemical with cyanide attached and found that it killed 100 percent of treated aphids even at the infinitesimal concentration of 2 parts per 100,000.[435] Within a few weeks, Schrader experimented with manufacturing techniques and discovered that his new chemical "exercised an extremely unpleasant toxic effect on man."[435] "The first symptom noticed," Schrader recollected, "was an inexplicable action causing the power of sight to be much weakened in artificial light. In the darkness of early January it was hardly possible to read by electric light, or after working hours to reach my home by car."[435] Schrader noted that "the smallest quantity of the substance VII dropped by inattention on the bench, caused strong irritation of the cornea, and a very strong feeling of oppression in the chest."[435]

Schrader recovered from this accidental exposure to the world's first organophosphate nerve gas. On February 5, 1937, Schrader sent a sample of his discovery to a professor at the Factory Hygiene Institute of Elberfeld.[434,435] In March, Schrader applied for patent protection of the chemical series to which his "substance VII" belonged, with the hope that the chemicals could be used as insecti-

Fig. 9.1. Gerhard Schrader in his laboratory at I. G. Farben.
From Bayer AG Corporate History and Archives.

cides.[435] A 1935 Nazi decree required any patent application with military potential to be kept strictly secret. By the time "German Pat. 155/39 (Top Secret)" was issued, the extreme toxicity of "substance VII" to mammals had been confirmed by Schrader's colleague via testing on mice, guinea pigs, rabbits, cats, dogs, and primates, and so died its potential commercial application as an insecticide.[435]

Schrader's colleague relayed the discovery to the Army Weapons Office, which immediately asserted control over research and production under the utmost secrecy. "I was a few days later requested," reported Schrader, ". . . to demonstrate the preparation of the cyanide (VII) in the Army Anti-Gas Laboratories, Spandau-Citadel, Berlin. Colonel Rüdiger, at that time head of the department concerned, recognised the significance of the new substance for military purposes. He arranged the remodeling of the chemical laboratories in Spandau and provided for the installing of a modern technical experimental station for the manufacture of the cyanide."[435]

The Nazis immediately discovered the benefits of Schrader's new chemical. It had no color and almost no odor, and it could kill via either inhalation or skin penetration. That same year, in 1937, Schrader relocated to a new I. G. Farben factory where he could "pursue the study of organic phosphorus compounds undisturbed."[435] He received a bonus of 50,000 Reichsmarks, sixteen times the annual salary of the average German worker.[362]

Schrader described the detailed order of events that followed.

In the years from 1937 to 1939 the H.W.A. [Heereswaffenamt, the German Army Weapons Agency] busied itself with the technical production of the cyanide (VII). This substance had been given the preparation number 9/91 by me. Prof. Gross called it "Le 100." The H.W.A. named it "Gelan," and later, "Substance 83." In the year 1939, the H.W.A. built its own plant for the production of substance 83 at Munster-lager (Heidkrug, Raubkammer). At the end of 1939, Herr Director Ambros received the order from the High Command to establish a special factory for the large scale production of substance 83. As a site for the new establishment a place near Dyhernfurth/Oder (about 40 Km. N.W. of Breslau) was chosen. The building of the new works began in Autumn 1940, and in April 1942, the Anorgana concern began the production of substance 83, which was then termed "Trilon 83" later "T.83" and, finally, "Tabun."[435]

Hermann Göring served as one of the first Nazi ministers; created the Nazi secret state police force, the Gestapo; served as commander in chief of the Luftwaffe and plenipotentiary of the Nazi Four Year Plan; and rose to the number two position in the Nazi command after Hitler. After the war, Göring was also one of the major war criminals sentenced to death by the Nuremberg judges, and he requested that his execution take place by firing squad rather than hanging.[362] The Allied Control Commission denied his request, and Göring killed himself with potassium cyanide that he had hid-

Fig. 9.2. The I. G. Farben tabun factory in Dyhernfurth, Lower Silesian Province, Poland, photographed in 1941 by a British spy plane (upper right portion of photo). Otto Ambros designed and managed the secret facility.[362] Here, three thousand slaves filled artillery shells and bomb casings with tabun. Here, too, the reliability of gas masks developed by SS-Brigadeführer and chemist Walter Schieber were tested on concentration camp prisoners who were sprayed with nerve gases. Soviet forces captured Dyhernfurth on February 5, 1945. By then, the SS guards had marched the slaves to the Gross-Rosen concentration camp (one-third of them survived the march), the munitions had been secreted away, documentary evidence had been destroyed, and I. G. Farben employees had bolted. A Nazi platoon shelled the Soviet forces as a diversion to allow an I. G. Farben technical team to scrub clean the tabun facility. When the Soviets discovered the facility, it was empty of people and tabun, but they were able to disassemble the factory and ship it home for reassembly and tabun production in a new facility outside Stalingrad. © HES. From National Collection of Aerial Photography, NCAP-000-000-036-543, ncap.org.uk.

den for eighteen months in a vial that he kept alternately in his anus and his navel. During the war, Göring had a hand in diverse aspects of Nazi warfare, including the production of tabun. (The name "tabun" was derived from the word "taboo.")[362]

After tabun's discovery, Göring commissioned a report from the head of I. G. Farben's board of directors, Karl Krauch. Krauch wrote that tabun was "the weapon of superior intelligence and superior scientific-technological thinking," which could be "used against the enemy's hinterland."[362] Göring agreed with this assessment, and wrote to Krauch that nerve agents would induce "psychological havoc on civilian populations, driving them crazy with fear."[362] On August 22, 1938, Göring appointed Krauch as his plenipotentiary for special questions of chemical production, with a focus on tabun.[362]

Schrader's work progressed, and on December 10, 1938, he discovered a compound ten times more toxic than tabun.[435] "The action of this substance as a toxic war substance was," he reported, "in comparison with hitherto-known substances, astonishingly high."[435] Tests revealed that "the toxic value of this interesting substance against warm blooded animals surpassed that of Tabun, and that it could not be considered as an insecticide."[435]

Like tabun, the new substance carried a series of labels. Schrader referred to it as "Le 213" and reported that "the H.W.A. coded the substance as 'Stoff 146.' Later the preparation was called 'Trilon 146,' 'T.46' and finally 'Sarin.'"[435] "Sarin" was an acronym for his name and those of his colleagues who assisted in its development: Schrader and Ambros from I. G. Farben along with Rüdiger and van der LINde from the German army.[362,431]

In June 1939 Schrader's formula for sarin was ready for the Wehrmacht's laboratory in Berlin, which produced its first sarin samples in September.[431,435] Simultaneously, Germany crushed Poland, and Hitler announced in a fiery speech meant for Allied consumption that he possessed new weapons against which there was no defense.

Large-scale manufacture of tabun and sarin posed many challenges, both financial and technical. Germany constructed its tabun factory using Wehrmacht funds that were channeled through newly created companies with the intent to obfuscate the role of I. G. Farben.[431] Technical issues proved more obstinate, and problems in tabun synthesis quickly mounted. Tabun's ingredients were corrosive to steel and iron, which therefore had to be plated with silver. The toxicity of tabun posed special difficulties for the workforce and operations of the factory. Steam and ammonia were deployed to decontaminate equipment, and workers used respirators and wore rubber suits usable for only ten work shifts. Nevertheless, before the factory churned out its first batch of tabun, more than three hundred accidental exposures occurred. The worst cases caused death in two minutes. High body fat lowered the side effects of tabun exposure, so the factory provided workers with high fat content foods. Casualties among both workers and slave laborers provided Nazi scientists with data on the toxic effects of tabun on people.[436]

Once technical problems were solved, the factory had a production capacity of 1,000 metric tons of tabun per month.[437] First, the ingredients were produced and then synthesized into tabun, which was then transferred to an enormous basement facility for the filling of bombs and shells.[431] When ready for deployment, the tabun munitions were secreted offsite and stored in an underground arsenal in Upper Silesia.

The Nazis also admired the extra lethality afforded by sarin, which they created in a secret facility called Building 144.[431] The nerve agent research and development facilities employed 1,200 workers. Sarin and tabun were cleverly weaponized with a variety of delivery agents, including a machine gun that could fire 2,000 rounds of sarin- or tabun-filled bullets per minute.

By the early 1940s, Schrader had screened between one hundred and two hundred chemicals for toxicity as war gases for the Nazi government.[303] The Nazis tested the toxicity of nerve gases and other chemicals produced and supplied by I. G. Farben on inmates of concentration camps.[311,400,426] They created a museum to display the organs of animals gassed with tabun and four thousand photographs of people exposed accidentally or experimentally.[431]

Schrader also created a plethora of organophosphate insecticides, including parathion and malathion.[303] He discovered that parathion had greater toxicity for insects than DDT; unlike DDT, it killed every insect species against which it was applied. The discovery of these compounds and their toxicity for insects and humans occurred at the very time that the Nazi propaganda machine spewed out rhetoric equating Jews with insects and other pests that needed to be exterminated.[400]

The Nazis drew inspiration from German texts from the previous century that pronounced Jews "vermin, spiders, swarms of locusts, leeches, giant parasite growths, poisonous worms."[400] The famous nineteenth-century German biblical scholar Paul de Lagarde stated of the Jews: "One does not have dealings with pests and parasites: one does not rear them and cherish them; one destroys them as speedily as possible."[438] Hitler extended this description to call Jews a "pestilence" and "carriers of bacilli worse than Black Death."[400] Goebbels said, "Since the flea is not a pleasant animal we are not obliged to keep it, protect it and let it prosper so that it may prick and torture us, but our duty is rather to exterminate it. Likewise with the Jew."[400]

Eventually the Nazis reached a monthly production capacity of 12,000 tons of war gases, including tabun, varieties of mustard gas for both warm and cold conditions, and an incendiary gas called *N-Stoff* that could even burn asphalt.[431] The Luftwaffe stockpiled nearly half a million gas bombs, ranging in size from 15 kg to 750 kg, containing phosgene, hydrogen cyanide, mustard gas, and tabun, as well as a variety of other gases, acids, and bases.

Nazi leaders intensely felt the pressure to use this arsenal. General Hermann Ochsner, in charge of the Nazi chemical troops, stated his position at the onset of the war that these gases were a potent weapon of terror.[431] "There is no doubt," he said, "that a city like London would be plunged into a state of unbearable turmoil which would bring enormous pressure to bear on the enemy government."

By 1944, the Nazis were launching flying bombs (V-weapons) into England in batches of up to two hundred at a time.[431] Two thousand of these rockets rained down on Britain in the first two weeks of the offensive, which the Allies code-named "Crossbow." Fifty tons of V-weapons exploded in London daily, and the United Kingdom devoted half of its air power in a desperate attempt to shoot down the rockets. Although they were capable of delivering tabun in these rockets and by numerous other means, the Germans did not do so. But they clearly came close. Hitler told Mussolini before D-Day in 1944 that he possessed weapons that "would turn London into a garden of ruins."[431] Members of Hitler's inner circle, including Martin Bormann (Hitler's private secretary), Joseph Geobbels

(Reich minister of propaganda), and Robert Ley (head of the Nazi trade union), advocated for the unleashing of tabun.

The Allies were unaware of the existence of the German nerve gases. General Bernard Montgomery even left his antigas equipment behind in England when he landed his troops on the coast of Normandy.[431] The reason the D-Day invasion did not encounter gas attacks likely rested with a misinterpretation of Allied capabilities on the part of the Nazis.

Ambros worked with Schrader in the development of sarin, and he served as the director of the largest tabun production company. Therefore, his knowledge of German chemical weapons capabilities was second to none. Ambros and Albert Speer, who served as the minister of armaments and Hitler's architect, met with Hitler in May 1943 after the defeat at Stalingrad.[431] Nearly two years later, Speer plotted to assassinate Hitler by introducing tabun into the ventilation of Hitler's bunker, but he was unable to overcome technical pitfalls of employing tabun in such a manner.[326] After the war, he was one of the twenty-two accused major war criminals tried at Nuremberg, and he was the only one who pled guilty; the judges sentenced him to twenty-two years' imprisonment.

The agenda of the May 1943 meeting focused on the potential use of chemical weapons to reverse the Soviet advance. Ambros pointed out that the Allies had superior capacity to produce chemical weapons. Hitler admitted that was likely the case for the previous generations of gas weapons, then added, "but Germany has a special gas, Tabun. In this we have a monopoly in Germany."[426] Ambros provided the opposite, and incorrect, assessment. "I have justified reasons to assume," he said, "that Tabun, too, is known abroad . . . I am convinced that other countries, in case the German side might use these gases, would very shortly not only be able to imitate these special gases, but even produce them in much larger quantities."[426] This erroneous assessment, and its implication that Germany would incur more damage from retribution via gas weapons than it could induce by first use, may be the reason why the Nazis did not use their chemical arsenals on the battlefield.

During the war, American scientific journals ceased publication of articles about chemicals related to nerve gases.[431] The Nazis followed American technical publications, and they rightly assumed that the sudden halt to such publications resulted from US censorship. The science blocked from publication, however, was the experimental trials of DDT. Ironically, Müller's employer, J. R. Geigy, had already relayed the discovery of DDT to the Nazis in 1942, but the DDT secrecy on the part of the United States had the unintended consequence of convincing Nazi scientists that the Allies, too, had developed organophosphate nerve gases.[303,431]

The Allies also came within a breath of using their chemical arsenal. In June 1940, the chief of the Imperial General Staff, Sir John Dill, summarized his advocacy for first use of chemical weapons should German soldiers land on British soil. He wrote in a military memo, "At a time when our National existence is at stake, when we are threatened by an implacable enemy who himself recognizes no rules save those of expediency, we should not hesitate to adopt whatever means appear to offer the best chance of success."[431] One of Dill's own senior staffers disagreed and wrote that should the British be the first to use chemical weapons, "some of us would begin to wonder whether it really mattered which side won." Winston Churchill sided with Dill.

Despite more than two years of intense pressure from Churchill and the war cabinet, the British chemical weapons program was frustratingly slow to ramp up, and Churchill felt that this failure left Britain dangerously vulnerable to a German invasion.[431] "What is the explanation of the neglect to fulfill these orders," he wrote, "and who is responsible for it? . . . Those concerned should be beaten up."[431]

By July 1944, British stockpiles of chemical weapons were sufficient not only for home defense, but also for offense. Churchill wrote in a memo to his chiefs of staff that public opinions on which weapons and uses were considered moral vs. immoral shifted on a short time scale. "It is simply a question of fashion changing as she does between long and short skirts for women . . . I do not see why we should always have all the disadvantages of being the gentleman while they have all the advantages of being the cad . . . I quite agree that it may be several weeks or even months before I shall ask you to drench Germany with poison gas, and if we do it, let us do it one hundred percent. In the meanwhile, I want the matter studied in cold blood by sensible people and not by that particular set of psalm-singing uniformed defeatists which one runs across now here now there."[431]

The British prepared for chemical war not only through production of massive stockpiles of gas weapons, but also by manufacturing 70 million gas masks, 40 million containers of antigas ointment, and 40,000 tons of bleach to be used for decontamination.[431] By the end of the war, Allied and Axis nations had amassed about half a million tons of chemical weaponry, none of which was employed on the battlefield. The total mass of poison gas used in World War I amounted to only 20 percent of this World War II stockpile.

The only mass casualties caused by chemical weapons in the European Theater (excluding the use on concentration camp prisoners) were the result of a December 2, 1943, raid by the Luftwaffe on an Allied armada in the Italian port of Bari.[431] Among the destroyed ships was the American Liberty ship *SS John Harvey*, which had secretly carried a load of 2,000 mustard gas bombs into port. It was the worst Allied loss at sea since Pearl Harbor, and Churchill was appalled that the United States ordered the ship into the vulnerable Italian waters. General Eisenhower tried to keep the cause of the disaster a secret with strict censorship orders that were approved by President Roosevelt and the British War Cabinet, but the reason for the mass casualties among civilians and soldiers was impossible to contain. So the Combined Chiefs of Staff announced, "Allied policy is not (repeat not) to use gas unless or until the enemy does so first but that we are fully prepared to retaliate and do not deny the accident, which was a calculated risk."[431]

Nazi secrecy over tabun and sarin remained intact throughout the war. Scientists knew only the steps for their particular stage of the chemical synthesis, rather than the complete recipe.[431] Even Schrader did not have full access to the research he unleashed. The chemicals were discussed using their many pseudonyms, and even the ingredients for tabun and sarin were called by false names that repeatedly changed. The Nazis buried the ledger bearing this information at the end of the war.

A single leak nearly breached this wall of secrecy. In May 1943, the British military captured a German chemist in Tunisia.[431] The chemist spilled what he knew of Trilon 83 (tabun), "a clear colorless liquid with little smell" that "cannot be classed with any of the other war gases as it is a nerve poison" causing the pupils of the eyes to shrink "to a pinhead and asthma-like difficulties in breathing. In any heavier concentrations death occurs in about a quarter of an hour." The German chemist provided detailed information on ingredients, effects, and means of delivery and defense. British interrogators documented this information in a top-secret memo, which intelligence officials ignored even though the British had already tested chemicals with effects similar to those of tabun.

Organophosphates (1944–59)

The answer to a farmer's most fantastic dream, plants that kill their own insect pests, now appears within the realm of possibility. This dream lies in new systemic insecticides, phosphorus compounds developed in Germany, which are used not on the outside of the plant but inside it.

—*Science News Letter*, in an editorial about Schrader's systemic insecticides, 1951[439]

The shock of tabun's existence was only realized in April 1945, when British troops from Montgomery's Twenty-first Army Group captured an abandoned German army proving ground called Raubkammer, "the Robbers' Lair," as well as a nearby cluster of bunkers.[362] The Robbers' Lair contained a zoo for testing chemical weapons on animals, and the bunkers contained shells with an unknown substance. Chemical weapons specialists from the United States and United Kingdom descended upon the site. They field-tested the unknown substance on rabbits in a mobile lab and discovered its unparalleled toxicity. They carefully shipped the substance to the United Kingdom, where scientists at the Chemical Defence Establishment analyzed the contents. In a single weekend, and despite the constriction of their pupils caused by accidental exposure, the scientists discovered the composition and toxicity of tabun and the benefits of its antidote, atropine.[440]

With the war winding down in Europe, the US Chemical Warfare Service prioritized the importation of Hitler's chemists and their organophosphate nerve agents in order to advance US chemical weaponry.[362] The priority had shifted from conquering Germany to containing the Soviet Union and conquering Japan, and thus captured German chemists and laboratories were prized assets.[431] General William Porter, who led the Chemical Warfare Service, ordered the shipment from the Robbers' Lair to the United States of five of the 260 kg tabun bombs for field testing.[362] Within a few months, US forces had shuttled 530 tons of tabun to the United States for testing. The Chemical Warfare Service also began the process of bringing German chemists to the United States to assist with chemical weapons development, despite objections from the State Department.

Allied troops arrested Schrader in March 1945, following the occupation of Leverkusen.[434] He was held with other prominent German scientists at Kransberg Castle, a medieval castle situated in the Taunus Mountains that had served as the headquarters of the Luftwaffe.[362] The Allies code-named the castle "Dustbin," and there they held more than twenty I. G. Farben chemists and six board members, in addition to other Nazi scientists, physicians, and industrialists. At Dustbin, a team from the British Intelligence Objectives Sub-Committee (BIOS) interviewed Schrader in August–September 1945.

Schrader cooperated with Allied investigators and produced both a classified report on organophosphate nerve gases[435] and an unclassified report on organophosphate insecticides.[441] Schrader prepared the unclassified report at the request of the BIOS investigators to facilitate commercialization of his discoveries. The introduction to the unclassified report stated, "The information now reported has already been recorded in BIOS Reports, which, however, owing to their content of other data, have necessarily been graded 'Secret.' The following account, which deals solely with the insecticidal aspects of Schrader's work, has therefore been prepared, in order to give greater availability to the information . . . A word of warning to investigators in the field covered by this report appears to be necessary. Some of the compounds described, although primarily insecticidal are also toxic to warm-blooded animals. It is known that other substances of similar general type are still more toxic to higher animals, and it is quite possible that workers undertaking synthetic studies in this series may prepare a substance sufficiently toxic as to constitute a real danger to themselves and others in the vicinity."[441] BIOS investigators released Schrader from custody, and he returned to work, rebuilding his destroyed laboratory and improving upon his organophosphate insecticides.

Meanwhile, as Germany collapsed in the spring of 1945, tabun and sarin factories in the east fell under Soviet control.[431] The Soviets also found in their conquered territory an even more potent Nazi nerve agent called soman, discovered by the organic chemist Richard Kuhn and his collaborator Konrad Henkel.[436] Kuhn, like Ambros, completed his PhD under the supervision of the Jewish Nobel Laureate Richard Willstätter.[304] Despite his close connection to his Jewish mentor, when Kuhn won the 1938 Nobel Prize in Chemistry, he did not initially accept it because Hitler called it a Jewish prize.[362]

Kuhn synthesized soman in the summer of 1944 while studying the toxicology of tabun and sarin for the German army.[442] Kuhn and his researchers found that tabun and sarin were toxic because they inhibited an essential neurotransmitter, leading to rapid death.[436] They also discovered that atropine could be used as an antidote.

Schrader described the discovery of soman to his British interrogators: "In 1944 my work was made known by the H.W.A. to Prof. R. Kuhn without my knowledge. Kuhn introduced into the Sarin molecule pinakolyl alcohol in place of the isopropylalcohol. The product obtained was codified by the War Office as SOMAN. I produced this product in August/44 and investigated it. Against warm-blooded animals the action of Soman is perhaps twice as strong as with Sarin. For the protection of plants the Sarin-Soman series are of little importance on account of their very strong physiological action."[435] The Allies later called tabun "GA" for "German agent A"; sarin, "GB" for "German agent B"; and soman, "GD" for "German agent D."[315]

Kuhn discovered soman too late in the war for the Nazis to weaponize it. After the war, Allied investigators interrogated Kuhn.[362] Kuhn denied his participation in the Nazi chemical weapons program. One of his interrogators summed up the disbelief in Kuhn's claim. "Richard Kuhn's record did not seem too clean to me. As president of the German Chemical Society he had followed the Nazi cult and rites quite faithfully. He never failed to give the Hitler salute when starting his classes and to shout 'Siegheil' like a true Nazi leader."[362]

Indeed, Kuhn approved the funding for Otto Bickenbach's grant application entitled "Biological and Physical-chemical Experiments on Protein-plasma Substances Regarding the Effects of Chemical Warfare Agents and Bacterial Poisons."[436] Bickenbach used the funding for human experiments with the chemical weapon phosgene at the Natzweiler concentration camp. The Nazis also tested chemical weapons on concentration camp inmates at Sachsenhausen and Neuengamme.[443] Bickenbach's attorney at his war crimes trial after the war requested of Kuhn a statement for the defense, to which Kuhn provided, "I know Mr. Otto Bickenbach, who practiced in Heidelberg for many years, personally and scientifically. I do not doubt for a moment that his experiments with hexamethylene-tetramine were scientifically outstanding, precise and that he pursued a high goal for the whole of mankind. The heroic self-experiment, of which I have now learned, confirms me in my conviction. It is my opinion that the results achieved by him—and those one can expect from him—will contribute to the blessings of many."[436]

While the Soviets focused their research on developing soman, the British concentrated their experiments on sarin. British researchers conducted exposure trials on volunteer servicemen, as well as on chimpanzees, goats, dogs, and other mammals.[431] Despite casualties among their volunteers, the British soon produced 6 kg of sarin per hour. The Americans had a similar sarin production program, though at a much larger scale and at a cost of only three dollars per kilogram.

At the conclusion of the war in Europe, various US government agencies formalized their recruitment of Nazi scientists, doctors, and engineers for classified research and development of American technologies and weaponry in "Operation Overcast."[303] The Joint Chiefs of Staff approved the operation on July 6, 1945, in a classified memo whose subject read, "Exploitation of German Specialists in Science and Technology in the United States."[362] The Joint Chiefs did not initially notify President Harry S. Truman of the enterprise, though his approval was sought and obtained at the end of the summer of 1946.

The secret effort morphed into "Operation Paperclip" after German families compromised the original name by calling their military housing "Camp Overcast."[362] The new name derived from the practice of Army Intelligence officers, who would attach paperclips to the files of Nazis as a marker that the file should not be reviewed by the State Department because officials there would inevitably object to recruitment of Nazis. The qualifying language for recruiting German scientists, doctors, and engineers changed from allowing "no known or alleged war criminals" and "no active Nazis" to avoiding the recruitment of those who would "plan for the resurgence of German military potential."[362]

The program grew from modest to ambitious and from temporary visas to permanent residencies due to the emerging threat of the Soviet Union, and the resulting competition for the talents of German specialists.[362] Operation Paperclip recruited more than sixteen hundred Nazi scientists, doctors, and engineers into US agencies, one of which was the Chemical Corps (formerly the Chemical Warfare Service). One of the scientists recruited was Ambros, a recruitment undertaken by Lieutenant Colonel Philip R. Tarr, who served during the war as the chief officer in the US Army Intelligence Division of the Chemical Warfare Service, Europe.

Tarr also led the American contingent to the joint US–UK Combined Intelligence Objectives Subcommittee (CIOS), a group comprising more than three thousand technical experts tasked with translating and interpreting Nazi scientific documents as they were discovered, including those related to chemical weapons.[362] Tarr's British counterpart was Major Edmund Tilley, and Tarr kept Tilley in the dark regarding his efforts to recruit Ambros.

The Chemical Corps sought Ambros's detailed knowledge of production requirements of tabun and sarin, which they valued above justice for his crimes.[362] Immediately after the war, Tarr even dispatched Ambros without escort on a secret mission to obtain blueprints for the silver-lined machinery used to produce tabun. At Ambros's request, and in order to secure his cooperation, Tarr also tried to secure the release from Dustbin of all the Nazi chemical warfare scientists. Tarr went so far as to forge a command to that effect from the colonel in charge of the British Ministry of Supply. Ambros eventually slipped into the French zone, where he negotiated to exchange what he knew for a managerial job at an I. G. Farben factory. A British officer reported that Tarr had "taken steps to assist [Ambros] to evade arrest."[362] Tarr arranged for a Chemical Warfare Service inspector, on loan from Dow Chemical Company, to meet with Ambros in the French zone on July 28, 1945. Their meeting was productive, and the US chemist said to Ambros, "I would look forward after the conclusion of the peace treaty [to] continuing our relations [in my position] as a representative for Dow."[362]

Other officials in the US Army repeatedly tried to lure Ambros out of the French zone in order to execute an arrest.[362] Undercover agents of the Counter Intelligence Corps tracked Ambros's movements, but their ruse to trick Ambros into an ambush in the American zone failed because Ambros had his own agents tracking the American agents who tracked him, and he sent a double into the trap in his stead. The next day Ambros dispatched a note on I. G. Farben stationery to the American official who had set the trap; he wrote, "Sorry that I could not make the appointment."[362]

As Germany's losses mounted at the end of 1944, Ambros ordered his deputy, Jürgen von Klenck, to destroy all documents on war gases and contracts between I. G. Farben and the Wehrmacht.[362]

Unbeknownst to Ambros, von Klenck placed some of the key documents in a steel drum and had the drum buried on a farm. This document cache would prove to be Ambros's undoing.

Major Tilley discovered that Tarr had his own agenda that ran counter to their joint leadership of CIOS.[362] Tilley felt infuriated over Tarr's actions to protect Ambros and to use him for American chemical weapons development. Tilley finally caught a break when he discovered the hidden drum of I. G. Farben documents on October 27, 1945. The documents included proof of Ambros's guilt. Two days later, BIOS issued its own arrest warrant for Ambros.

Ambros remained safe in the French zone for another three months before he was captured on January 17, 1946, as he attempted to leave the zone. Major Tilley then interrogated Ambros at Dustbin before his transfer to the Nuremberg jail. Although convicted of crimes against humanity in Case VI of the Nuremberg trials, the case against I. G. Farben, Ambros received an early release from prison in 1951, which enabled him to work for the US Chemical Corps. Like many other ardent Nazis and war criminals, Ambros thrived in the newly competitive order of the Cold War.

Within a year of the war's end, the Chemical Corps had produced five organophosphates and improved the equipment for the efficient synthesis of sarin.[303] It tested these gases on insects, both to determine their efficacy as insecticides and to study the potential use of insects as biological indicators of nerve gases on the battlefield. Through a subcontract with the Johns Hopkins School of Medicine, the Chemical Corps also tested organophosphate insecticides on human volunteers in order to refine protocols for identifying and treating nerve gas poisoning.

The Chemical Corps thus used its wartime experience and expertise to expand its appeal to society by conducting experiments and field trials that ushered new pesticides into the marketplace.[303] It promoted its decontamination devices, mortars, and smoke generators as delivery tools for insecticides. Its flamethrowers and incendiary chemicals could be used to destroy weeds or melt ice and snow. Smoke generators could shield orchards from frost, and war gases could be employed for riot control. The potential applications of war gas technologies seemed endless, and these efforts created a constituency that ensured continued government funding for the Chemical Corps. At the same time, the US government subsidized chemical manufacturers via tax incentives, and the Department of Defense provided major contracts.

Research continued on the organophosphate family of insecticides. One, called amiton, developed in 1952 by the British company Imperial Chemical Industries (ICI) in its search for a new pesticide, proved highly toxic to spider mites, and so had great commercial potential.[434] However, farmers who used amiton in field trials experienced nerve impulse poisoning. The insecticide was so toxic that a few milligrams was sufficient to kill a person.[303,431] ICI realized that it could not safely market the chemical, so it provided a sample to British authorities, who relayed the discovery to the US Chemical Corps.

The Chemical Defence Experimental Establishment at Porton Down in the United Kingdom tested the potential of amiton as a chemical weapon. It found that amiton, more so than the G agents developed by Schrader and Kuhn, readily induced its toxic effects through the skin.[434] A chemist at Porton Down modified the structure of amiton and created the Venomous agents, or V agents. V agents applied to the skin had a toxicity a thousand times greater than sarin, and exposure of the skin to a drop the size of a pinhead was sufficient to kill in fifteen minutes.[315]

The most important chemical weapon to arise from the V agents was VX. Unlike the German nerve agents, which tended to evaporate, this new chemical was heavy and viscous, and could therefore render a battlefield uninhabitable for a considerable period of time.[431] In 1956, the United States and Britain collaborated to develop an efficient manufacturing process for VX, and the US production facility became operational in 1959.[303,431] Therefore, over the course of the 1950s, a commercial enterprise to develop a new pesticide evolved into the most toxic chemical weapon ever created, up to that point. Delivery systems enabled with VX included land mines, artillery, spray tanks, and missiles.[431]

The next innovative development was the binary weapon, in which the ingredients of nerve agents were stored separately within a rocket or shell; when the weapon was fired, a separation wall between the ingredients burst open, and the chemical reaction to create the nerve agent occurred in flight.[431] This technology averted the risks of working with the nerve agent directly; instead, the relatively innocuous ingredients were handled by people, while the delivery vehicle handled the reaction leading to the chemical weapon.

During the same postwar period when scientists and engineers improved upon organophosphate chemical weapons, other scientists working for pesticide manufacturers improved upon organophosphate pesticides. Schrader's unclassified report on insecticides contained gaps because I. G. Farben officials destroyed many of Schrader's records, and others were lost.[441] But the details were sufficient for English chemical companies to begin marketing parathion in February 1947.[434] Even before this, American chemical companies used Schrader's unclassified report to manufacture organophosphate pesticides without patent or license constraints.

Monsanto manufactured the first of these, known as HETP (also called TEPP), beginning in 1946 for sale as a rodenticide.[303] The Chemical Corps reported that 1 pound of the compound was sufficient to kill 2 million rats. Other organophosphate pesticides, produced by the chemical companies Hercules, American Cyanamid, Shell, Niagara, Stauffer, Chemagro, Victor, and Velsicol, quickly followed.

One of Schrader's most important discoveries, which he made during the war, was that some of his organophosphate pesticides passed into a plant's roots and then throughout the stem and leaves, making them "systemic pesticides."[303,435,444] When insect pests ate crops treated with them, they ingested the poison and died. Schrader initially discovered this startling "chemotherapeutic agent for living plants" while investigating insecticides based upon fluoroethyl alcohol.[435] He conducted trials in which he watered maize plants with a

0.1 percent solution of the insecticide. Eight days later he fed the stem and leaves to rabbits, which died within twenty-four hours. Similarly, caterpillars (devouring insects) and aphids (sucking insects) died after feeding upon the leaves. "This trial proved beyond a doubt," wrote Schrader, "that the substance XLVII is taken up by the leaves, as well as the roots. It is clearly present in the sap-stream since it reaches every part of the plant making it for a time toxic to insects, and in certain cases also to warm blooded animals. With this discovery, a problem which has been worked on for decades nears its solution."[435]

A singular advantage of this approach was the destruction of insects on parts of the plant inaccessible to an insecticidal spray solution, as the plant's circulatory system was the delivery agent of the poison. Many of Schrader's systemic insecticides protected plants from insect pests for considerable periods of time.

Schrader immediately employed this new technique to battle a *Phylloxera* infestation of German vineyards.[435] Up until Schrader's discovery, *Phylloxera* eradication had entailed barring entry to a large area surrounding the infestation, where pesticide applicators coated the plants with carbon disulphide. This resulted in the destruction of the vineyard and rendered the soil barren for four years, a disastrous result for the vineyard owner. With Schrader's technique, the vines remained standing and produced grapes, though sometimes the grapes also imbibed the poison. Even in that worst-case scenario, the vines produced healthy grapes on the next harvest. Similarly, Schrader's discovery of systemic pesticides opened new avenues for the destruction of insect pests of a wide variety of crops, and the technique met with great enthusiasm throughout the world.

Schrader also employed his skills to battle the Colorado potato beetle. Food was scarce in Germany during and after the war, and the potato beetle menaced the all-important potato crop in the summer of 1944 and immediately after the war in 1945. Schrader devoted himself to its destruction, and in 1945 he discovered new organophosphate insecticides that killed the potato beetle at a lower concentration and much more effectively than the insecticide they replaced, arsenate of lime. "Whereas arsenate lime action only kills off the potato beetle and its larvae;" wrote Schrader, "the new substance also kills with certainty the eggs laid on the under side of the plants. By this means a really efficient method of destruction of the potato beetle plague is a possibility."[435]

Schrader displayed no interest in chemical weapons after the war. But he worked passionately to develop better insecticides. Bayer AG, Schrader's employer and one of the chemical giants that arose from the ashes of I. G. Farben after the war, brought these chemicals to market, including systox in 1952, metasystox in 1954, and then dipterex, gusathion, and folidol.[434] Many other chemists synthesized organophosphate chemicals to discover new insecticides; by 1959, about fifty thousand such chemicals had been synthesized and tested for their ability to kill insects, and forty had been commercialized.[445]

During the war, Schrader's acclaim came only from the Nazi leadership; he was, otherwise, an unknown man since his many chemical discoveries remained secret. After the war, however, Schrader received widespread approbation for bringing potent insecticides to bear against man's insect enemies. He received awards from the German government and the Society of German Chemists, as well as from his employer.[434] He retired from managing the Bayer plant protection laboratory in 1967, and died in 1990 at the age of eighty-seven.

The organophosphate family of insecticides initiated by Schrader just before World War II competed in peacetime against the organochlorine family of insecticides initiated by Paul Müller at the outset of the war. Schrader's family of chemicals would eventually win this competition, though no one predicted the stimulus for the victory: a small book, written by a gentle woman who felt stunned by the environmental damage induced by pesticides, especially DDT.

PART 4

Ecology

[**10**]

Resistance

(1945-1962)

Have you walked up and down upon the earth lately? I have; and I have examined Man's wonderful inventions. And I tell you that in the arts of life man invents nothing; but in the arts of death he outdoes Nature herself, and produces by chemistry and machinery all the slaughter of plague, pestilence, and famine.—**The Devil speaking to Don Juan in Hell in the play** *Man and Superman* **by George Bernard Shaw**, 1903[446]

The eggs lie cold, the fires of life that flickered for a few days now extinguished.—**Rachel Carson**, 1962[447]

After World War I, the conquest of nature seemed prudent to many experts concerned about food security and the risk of vector-borne diseases. The chief entomologist for the US Department of Agriculture, Leland Ossian Howard, wrote succinctly of this challenge in 1922: "Few people realize the critical situation which exists at the present time. Men and nations have always struggled among themselves. War has seemed to be a necessity growing out of the ambition of the human race. It is too much, perhaps, to hope that the lesson which the world has recently learned in the years 1914 to 1918 will be strong enough to prevent the recurrence of international war; but, at all events, there is a war, not among human beings, but between all humanity and certain forces that are arrayed against it."[448]

Chief among these forces was the insect. With most organisms, Howard wrote, we feel "a certain sense of terrestrial fraternity."[448] Not so with the insect, which "brings with him something that does not seem to belong to the customs, the morale, the psychology of our globe. One would say that it comes from another planet, more monstrous, more energetic, more insensate, more atrocious, more infernal than ours."[448] The odds in this war seemed stacked against humanity, given that the insect was "so incomparably better armed, better equipped than ourselves, these compressions of energy and activity which are our most mysterious enemies, our rivals in these latter hours, and perhaps our successors."[448]

Howard arrived at these views through his long career battling insects.[449] Walter Reed consulted Howard before engaging in his investigation of mosquitoes as the possible vector of yellow fever in Cuba. Additionally, Howard published a book in 1901 that served as Major William Gorgas's instruction manual for the eradication of mosquitoes in Havana and the Panama Canal Zone: *Mosquitoes, How They Live; How They Carry Disease; How They Are Classified; How They May Be Destroyed*.[450] But the problem in the interwar years and earlier was the lack of effective chemicals with which to battle man's insect enemies.

World War II changed all that. No longer did humanity face down its insect foes armed only with arsenates, lime and sulfur sprays, plant extracts, and petroleum emulsions. Organic chemistry had burst onto the scene, and the list of new and effective chemical insecticides tipped the scale in humanity's favor. Entomologists wielded these tools with enthusiasm and used the language of war in their calls to action against insects.[303]

But not everyone felt sanguine about the wholesale use of chemicals against insects. Even when DDT was first used during the war, some proponents preached caution. Brigadier General James Stevens Simmons, a staunch advocate of DDT, wrote, "It is fully realized that such a powerful insecticide may be a double-edged sword, and that its unintelligent use might eliminate certain valuable insects essential to agriculture and horticulture. Even more important, it might conceivably disturb vital balances in the animal and plant kingdoms and thus upset various fundamental biological cycles."[349] The problems that Simmons warned about, related to the indiscriminate and pedestrian uses of DDT and other new pesticides, arose immediately after World War II, and a handful of scientists and professional public health personnel wrote of their concerns.

The scientists in charge of the US Communicable Disease Center (later changed to the Centers for Disease Control and Prevention, or CDC) wrote in 1948, "During 1946, and especially in 1947, there were widespread complaints that DDT was less effective than it had been during the first year of general use [1945] . . . It is believed that many of the complaints were largely psychological, deriving from a decreased tolerance to even small numbers of flies caused by a year or more of relative freedom from large numbers of flies as a result of the early applications of DDT."[451] Despite attributing public misgivings to shifting expectations, even these scientific leaders expressed concerns about the effects of pesticides on wildlife. "As long as insecticides are used only in domestic surroundings," they wrote, "there is no danger of creating undesirable biologic imbalances. However, when these chemicals are applied over extensive, unoccupied areas . . . serious consideration must be given to the hazard they may present to wildlife species related to the economy and happiness of man. The peril to these forms may be obvious, direct, and immediate, or it may be insidious, indirect, and delayed."[451] Similarly, a British malariologist advocated for DDT use against mosquitoes, but said, "DDT is such a crude and powerful weapon that I cannot help regarding the routine use of this material from the air with anything but horror and aversion."[303]

The CDC scientists also noted that the chronic toxicity of these new pesticides to people "is entirely unexplored."[451] Only acute cases of DDT poisoning of humans had been reported, with similar symptoms to those experienced by insects and birds: "Vomiting, numbness, and partial paralysis of the extremities, mild convulsions, loss of proprioception and vibratory sensation in the extremities, and hyperactive knee-jerk reflexes were the immediate toxic effects."[451] Just two years into civilian use of DDT, deaths from misuse mounted. "Perhaps the most complete account of a fatal poisoning involving a DDT formulation," wrote the CDC scientists, "is that of a 58 year old man who drank 120 ml. of a commercial preparation containing DDT, 5 per cent; 'Lethane' (384 special), 2 per cent; xylene, 7 per cent; and deodorized kerosene, 86 per cent, following it with a quart of milk and several glasses of beer. The symptoms consisted of the rapid onset of epigastric pain and the vomiting of bloody material. These continued with varying intensity until death in coma on the 7th day."[451]

One of the arguments raised to counter concerns for human and animal health was pesticide specificity. "Those who have been concerned with the development of this type of weapon for combating our horde of insect foes," wrote a prominent expert in 1946, "have long since learned that insecticides are decidedly specific—what will bring the agony of death to one insect in a few minutes is laughed off by another."[373] Yet this author admitted in the same article, "Just how DDT kills insects has not been determined."

DDT and other organochlorine insecticides were strikingly effective over long periods of time. "One of the outstanding characteristics of DDT is its persistence," wrote an expert.[373] But that persistence led to its own set of problems. The chief entomologist of a DDT manufacturer noted, "I also have talked with service men who witnessed a pre-invasion spraying of one mosquito-ridden island of the South Pacific by low-flying bombers, and was told that even after that single spraying the only insect left alive was one lone butterfly. However, on the basis of that very testimony it is my belief that DDT should be and will be put in the hands of trained users after the war. DDT kills a greater variety of insects than any other insecticide because of the long-lasting toxicity of its residual deposits to succeeding generations of insects. And therein lies its greatest danger. Mankind has many friends in the insect world and DDT is equally destructive to friend and foe. Should nature's balance in the insect world be upset, the effect on man would be deplorable indeed."[358]

Disruption of that ecological balance was noted in 1951, when researchers reported that DDT use caused an increase in citrus-damaging insects because of the loss of their insect predators.[452] The evolution of resistance to DDT was noted even earlier; in fact, the idea of insects evolving resistance to insecticides was first proposed in 1914,[453] and demonstrated in 1916 for hydrocyanic acid,[454,455] long before the advent of DDT, though this hypothesis received little attention.[303]

In Greece, DDT-resistant houseflies appeared in 1947, only one year after countrywide spraying began, and DDT-resistant mosquitoes, fleas, bedbugs, and cockroaches replaced their vulnerable compatriots over the next few years.[456] Lice, flies, and mosquitoes, the carriers of so many deadly diseases, were found to have evolved resistance to DDT and other new insecticides in many parts of the world, including the United States, by 1952.[303,457] For example, following the introduction of DDT to southern California in 1946, it took only two years for houseflies to evolve resistance, and then only two more years for houseflies to also evolve resistance to the replacement insecticides methoxychlor, lindane, chlordane, toxaphene, aldrin, and dieldrin.[457] The US Army found lice to be resistant to DDT in the Korean War Theater, and had to revert to the old solution of pyrethrum.

With resistance to insecticides now a serious military concern, significant resources were directed to research. The first prominent result was a National Research Council conference in 1951 in which the eminent evolutionary biologist Theodosius Dobzhansky explained that resistance was an inevitable outcome of evolution by natural selection, which had been thoroughly explained a century earlier by Charles Darwin. "Biologists should abandon the habit of thought," he stated, "inherited from the pre-evolutionary era, which regards each species, race, or population as an embodiment of a certain 'type' or 'norm.' "[457]

To counter the evolution of resistance, chemical companies concluded that society "must depend on an endless development of new insecticides of greater variety and specification."[303] Therefore, the evolution of resistance created a market for a steady stream of new insecticides. A series of organophosphates were introduced to the marketplace. American Cyanamid advertised that malathion "even kills flies resistant to DDT and other chlorinated hydrocarbon insecticides."[303] When insects then evolved resistance to organophosphates, chemical companies introduced a new family of insecticides, the carbamates. Union Carbide announced in 1957 that its new carbamate insecticide, Sevin or carbaryl, was "safe, inexpensive, stable, and of relatively broad spectrum effectiveness."[303]

By 1962, scientists had found that approximately one hundred forty species of insect pests had evolved resistance to DDT.[425] "The more you use it," said one expert, "the more you need to use it. We have created a more serious pest problem than we ever had before."[320] "Tragically," wrote a wildlife expert, "the obdurate insistence on using it to the bitter end of its effectiveness—mostly for private pecuniary advantage and in disregard of the social costs it imposes—has poisoned the world and impoverished the world's fauna."[458]

Damage to wildlife was originally discovered in South Pacific islands following aerial spraying of DDT during the war.[303] One entomologist wrote, "Saipan Island is approaching a condition of devastation . . . No birds, no mammals, no insects, except a few flies."[459] The nature writer Edwin Way Teale warned toward the end of the war against the wholesale use of DDT to wipe out insects once the war was over. He wrote, "Given sufficient insecticide, airplanes and lackwit officials after the war, and we will be off with yelps of joy on a crusade against all the insects . . . Dusting a field or wood from the air will have all the judicious foresight of machine-gunning a throng of friends in order to kill a fleeing bandit." The result, he wrote, would be "a drab and dreary monument—to man's unteachable folly."[460]

Just a year after the war, the National Audubon Society warned Americans of the environmental hazards associated with the mass spraying of DDT,[458] and the US Fish and Wildlife Service published a study on wildlife mortality due to DDT exposure.[461] In addition to the obvious cases of acute poisoning of wildlife, especially birds and fish, delayed effects due to environmental persistence and the chronic toxicity of DDT were discovered in the 1950s.[320,420] For example, DDT sprayed on elms killed robins that visited the trees the following year.[462]

DDT and other organochlorine pesticides, because of their high solubility in fat and persistence in the environment, magnified in concentration as they moved from prey to predator. For example, in 1948, in an effort to kill gnats, the Lake County Mosquito Abatement District applied DDD (an insecticide that is a metabolite of DDT) to the water of Clear Lake, California, at a concentration of 14 parts per billion.[320] It repeated the application using 20 parts per billion DDD in 1954 and 1957. This resulted in a concentration of DDD in plankton that was 265 times greater than the water concentration. Plankton-eating fish accumulated twice the DDD concentration of the plankton, and fish and birds that ate the herbivorous fish accumulated DDD to a concentration 85,000 times greater than the water concentration. Breeding failure and mass mortality of western grebes ensued. Similar breeding failures and population declines of birds, especially high trophic level birds, occurred throughout the world in areas with intensive use of organochlorine pesticides.[463] Included in the list of species in serious decline was the national emblem of the United States, the bald eagle, and the fastest animal on earth, the peregrine falcon. Birds were particularly vulnerable to DDT and its metabolites due to consequent eggshell thinning and reproductive failure.

Incredible damage to wildlife occurred in the monumental federal spraying programs, such as the one to control the fire ant across 27 million acres in nine southern states.[464] The US Department of Agriculture mounted a public relations campaign to convince Americans of the menace of fire ants and the untold damage they do to crops and livestock. The department then initiated the fire ant extermination campaign in 1957.[320,420] In reality, the fire ant was merely a nuisance, leading a wildlife biologist to state, "It's like scalping yourself to cure dandruff."[464] The *Saturday Evening Post* wrote, "The fire ant is undoubtedly a miserable pest to have around, but it is doubtful that the program justifies the slaughter of millions of birds, fish and small game."[465] Indeed, fire ants lived in nests, which made them vulnerable to spot spraying, and yet the Department of Agriculture broadcast dieldrin pellets over a vast landscape of southern states.[320]

In 1959, *Reader's Digest*, which had the largest circulation of any magazine in the United States, warned its readers of the dangers of broad-scale pesticide applications. "The United States is engaged in an intensive war against destructive insects. The weapons employed are powerful and widespread, and so is the controversy they have engendered. Billions of pounds of poisons were broadcast over 100 million acres of cropland and forest. More billions of pounds are being spread across the nation this year—against spruce budworm in northern forests, grasshoppers in nine million acres of wheatland in the Midwest, white fringed beetle in the Southeast; against sand flies, gnats, Japanese beetles, corn borers and gypsy moths."[464] The same article quoted a prominent zoologist: "The current widespread program poses the greatest threat that animal life in North America has ever faced—worse than deforestation, worse than illegal shooting, worse than drainage, drought, oil pollution; possibly worse than all these decimating factors combined."[464] And the article also noted the possibility of growing dependency on chemical pesticides: "Since pesticides kill mice-eating hawks, owls and foxes as well as rodents, and beneficial as well as harmful insects, may we not find ourselves without natural allies in the war on pests, and become wholly dependent on ever stronger chemicals?"[464] Given the rapid evolution of resistance to pesticides, the article asked, "Are we trading a costly temporary victory over other pests for disaster in the form of superinsects later on?"[464]

Two years later, in 1961, *Reader's Digest* took the opposite approach in an article, "Good-By to Garden Pests?" The piece celebrated DDT, lindane, chlordane, malathion, and other prominent new pesticides, along with the contributions made by DuPont, American

Cyanamid, Velsicol, Esso, Shell, Dow, and Union Carbide to fulfill "the lazy homeowner's ultimate dream of a garden as a bit of paradise that will take care of itself."[466]

Citizen outcry over the negative effects of pesticides occurred sporadically throughout the United States, and was typically treated with indignation by government officials. The first major coalescing of public furor occurred in 1957 in reaction to the US Department of Agriculture aerial spraying of DDT to kill gypsy moths in southern New York.[420,464] In addition to spraying rural areas across 3 million acres, the agency sprayed urban communities in Westchester County and Long Island. A fog of DDT dissolved in kerosene fell upon commuters at train stations, children in playgrounds, housewives tending organic gardens—and often they were sprayed multiple times.[447] This led to an unsuccessful court case to halt the federal aerial spray program. The judge ruled that "mass spraying has a reasonable relation to the public objective of combating the evil of the gypsy moth and thus is within the proper exercise of the police power by the designated officials."[303]

These problems associated with the indiscriminate use of pesticides did not penetrate a broad audience. That required a cultural shift and the appearance of a voice that could synthesize a difficult topic into an engaging message.

The voice that galvanized public concern about the environment and sparked a fierce industrial backlash was that of the "gentle subversive," Rachel Carson.[467] Carson published her first story at the age of eleven, in 1918. Writing was forever her passion. That passion coupled itself to a fascination with the sea, which she had never seen, when, while a student at the Pennsylvania College for Women and during a thunderstorm, she read the last lines of Tennyson's poem "Locksley Hall":[468]

Cramming all the blast before it, in its breast a thunderbolt.
Let it fall on Locksley Hall, with rain or hail, or fire or snow;
For the mighty wind arises, roaring seaward, and I go.

Coincidentally, at about the same time that Carson read "Locksley Hall," Winston Churchill declared it the most prescient writing ever penned.[327] It certainly proved so for Carson. After graduating from college in 1929, Carson realized her dream to study for a summer at the Marine Biology Laboratory at Woods Hole in Massachusetts, and then to enroll in the master's program in biology at Johns Hopkins.[467]

But Carson's momentum faltered during her first semester, when the stock market crashed.[467] Her graduate student stipend totaled $200 per year. Her modest income suddenly became the mainstay of her financially stressed family, and her parents, her brother, her sister (a single mother), and her sister's two daughters all moved in with her in a house in Baltimore. Her father died, and then her sister also passed away, leaving behind the two girls. Carson, then twenty-nine years old, and her elderly mother assumed full responsibility for the children.

To support her small extended family in the midst of the Depression, Carson wrote reports and brochures on the sea for the US Bureau of Fisheries.[467] One of the brochures, her boss told her, was too good for that purpose, and he advised that she submit it to the *Atlantic Monthly*. How fortunate she was to have such a boss, because she found immediate success and received a $100 payment for this short piece entitled "Undersea," published in the *Atlantic* in 1937.[469] The writing was so good that an editor from Simon & Schuster and an important nonfiction writer encouraged her to expand the four pages into a book.[467]

Between family obligations and work for the federal government, the writing proceeded slowly, but Simon & Schuster published *Under the Sea-Wind: A Naturalist's Picture of Ocean Life* in November 1941.[470] The beautiful writing received instant praise from reviewers, but the timing of the book could not have been worse.[467] A month after its release, her book about sea creatures competed for attention with news of the Japanese attack on Pearl Harbor. The book was not a commercial success. Even the material necessities of war conspired against the book's sales, as the planned British edition fell to the reality of a paper shortage. By the end of the book's five years of active shelf life in stores, Carson had earned less than $700 in royalties.

Carson continued her writing duties for the federal government, and rose in the ranks of what was now the US Fish and Wildlife Service.[467] Her position as a writer for the government agency most concerned with animal conservation gave her a front-row seat to the budding concerns about DDT that emerged with its public release in 1945. The service tasked Carson with editing reports on the dangers of DDT, and thus she was one of the first people made aware of these risks.

"We have all heard a lot about what DDT will soon do for us by wiping out insect pests," she wrote to the editors of *Reader's Digest* in her proposal to write about DDT for the magazine. But she also noted that researchers were examining "what it will do to insects that are beneficial or even essential; how it may affect waterfowl, or birds that depend on insect food; whether it may upset the whole delicate balance of nature if unwisely used."[471] She was, perhaps, influenced by Edwin Way Teale, who also wrote in 1945, "Those who, today, would remold the world nearer to their hearts' desire with DDT, visualize their paradise as a realm devoid of this insect or that insect. They cherish an old, old illusion. They still imagine—in spite of a thousand bitter lessons to the contrary—that they can pull out threads here and there from the fabric of Nature without otherwise changing the web."[460] Carson's proposal to write a popular account of DDT for *Reader's Digest* met the same fate as so many other writing proposals about science for the general public. As a result, Carson moved on to other topics.

In 1951, ten years after publishing her first book on the sea, Carson published her second, *The Sea around Us*.[472] This time the stars aligned, and the book experienced startling success.[467] "Great poets from Homer down to Masefield," wrote the *New York Times* in a review, "have tried to evoke the deep mystery and endless fascination of the ocean. But the slender, gentle Miss Carson seems to have the best of it. Once or twice in a generation does the world get a physical scientist with literary genius. Miss Carson has written a classic in *The Sea around Us*."[463] Another *New York Times* review closed with this lament: "It's a pity that the book's publishers did not print on its jacket a photograph of Miss Carson. It would be pleasant to know what a woman looks like who can write about an exacting science with such beauty and precision."[473] The *Boston Globe* wrote, "Would you imagine a woman who has written about the seven seas and their wonders to be a hearty physical type? Not Miss Carson. She is small and slender, with chestnut hair and eyes whose color has something of both the green and blue of sea water. She is trim and feminine, wears a soft pink nail polish and uses lipstick and powder expertly, but sparingly."[474]

The *New Yorker* condensed the book into a series entitled "Profile of the Sea," for which it paid Carson $7,200.[467] The *Yale Review* paid her $75 to publish a chapter, and this led to a $1,000 prize from the American Association for the Advancement of Science. Book-of-the-Month Club selected *The Sea around Us* as an alternate, and *Reader's Digest* published an abridged version.[471] Carson received a Guggenheim Fellowship, and the *Saturday Review of Literature* profiled Carson. Accolades poured in, including the National Book Award for nonfiction. The book sat atop the *New York Times* bestseller list for a record 86 weeks, was published in 32 languages, and sold over 1.3 million copies in the first edition. Carson was forty-four years old, and success had found her at last. She could now devote her professional life entirely to writing.

At her acceptance speech for the National Book Award, Carson gave a hint of her writing to come. She wondered "if we have not too long been looking through the wrong end of the telescope. We have looked first at man with his vanities and greed, and at his problems of a day or a year; and then only, and from this biased point of view, we have looked outward at the earth and at the universe of which our earth is so minute a part. Yet these are the great realities, and against them we see our human problems in a new perspective. Perhaps if we reversed the telescope and looked at man down these long vistas, we should find less time and inclination to plan for our own destruction."[471]

The success of *The Sea around Us* also led to the reissue in 1952 of *Under the Sea Wind*, which quickly joined *The Sea around Us* on the bestseller lists.[471] The *New York Times* stated that these events were a "publishing phenomenon rare as a total solar eclipse."[463]

The Edge of the Sea followed in 1955.[475] "You have done it again!" wrote Edwin Way Teale when the *New Yorker* published the first serial installment.[471] The new book also received widespread praise and awards, and found its place on the *New York Times* bestseller list. That paper's book reviewer commented that "her understanding is so contagious that you find yourself taking an enormously friendly interest in all sorts of spiny and slime-wreathed creatures that you've hitherto regarded with hearty loathing."[476] Carson's ability to engage the reader on such loathsome subjects was key to her most important, and final, effort of all, the engagement of the reader on the subject of pesticides.

The subject was one in which she first tried to interest the children's author E. B. White, who wrote on many topics, including the environment, for the *New Yorker*.[463] White turned the suggestion back on her, proposing that she be the one to write about pesticides. "I think the whole vast subject of pollution," he wrote to her in a letter, "of which this gypsy moth business is just a small part, is of the utmost interest and concern to everybody. It starts in the kitchen and extends to Jupiter and Mars. Always some special group or interest is represented, never the earth itself."[463]

In 1957, domestic stresses once again plagued Carson and intruded upon her ability to write.[471] Her niece died, leaving behind a five-year-old boy born out of wedlock. Carson, at forty-nine years old, now cared for both her elderly mother and her young and often sick grandnephew. Nevertheless, the following year Carson's writing project on pesticides began to take shape. She gathered information amid feelings of horror at the indiscriminate use of pesticides by the US government against the fire ant and gypsy moth, and the disregard for unintentional effects these chemicals could have on people and wildlife.

Carson was fully cognizant of the stifling political climate of 1950s America, the paranoia about the communist menace, the commanding respect afforded to government authority, and the displays of religious and patriotic fervor.[467] As she started to piece together her argument on pesticides, an author she esteemed, John Kenneth Galbraith, published his critique of consumerism, in which he summarized the cultural challenge that lay before Carson: "These are the days when men of all social disciplines and political faiths seek the comfortable and the accepted; when the man of controversy is looked upon as a disturbing influence; when originality is taken to be a mark of instability; and when, in minor modification of the scriptural parable, the bland lead the bland."[477] So much more so, Carson must have thought, for the woman of controversy and originality.

But in the years leading up to the publication of *Silent Spring*, which had a working title of "Man against the Earth," events paved the way for her message.[466,467,471] In 1954, a Japanese fishing boat called the *Lucky Dragon* passed through radioactive fallout from the testing of an American hydrogen bomb on Bikini atoll.[478] The sick and dead crew members, struck down through their innocent wanderings in search of tuna, alarmed the world about the deadly dangers of radiation. In 1957, the Soviet Union mastered the technology of the intercontinental ballistic missile capable of carrying nuclear warheads.[425] Americans were also horrified to learn that strontium-90 from fallout, which has a half-life of nearly twenty-nine years, could appear in cow's milk and then incorporate into the bones and teeth of their children, potentially causing cancer.[479] Indeed, strontium-90 was documented in babies' teeth in 1961.[480] The National Committee for a Sane Nuclear Policy rallied people against nuclear testing and "extermination without representation."[467,481] Housewives organized the Women's Strike for Peace with the goal to "end to the arms race, not the human race."[482] And the Soviet Union stationed nuclear missiles in Cuba.[483]

It was not only enlightenment about radioactive fallout that challenged the public complacency of 1950s America. It was also chemical scares. The dangers of pesticides polluting the food supply gained traction in the media and in the public conscience during the Thanksgiving holiday of 1959, when the US Food and Drug Administration banned cranberries, which were already on store shelves, that had been treated with the herbicide aminotriazole.[484] The previous year, American Cyanamid had applied for a determination of allowable residue on food, but because aminotriazole caused thyroid cancer in rats, the Food and Drug Administration rejected the request. Cranberry growers, faced with heavy financial losses in a year with a bumper crop, demanded the resignation of Arthur Flemming, the secretary of the Department of Health, Education and Welfare, for his statement on the contamination of the crop. Growers wrote, "In justice to thousands of cranberry growers and distributors and millions of consumers, we demand that you take immediate steps to rectify the incalculable damages caused by your ill-informed and ill-advised press statement yesterday. You are killing a thoroughbred in order to destroy a single flea."[484] This incident awakened Americans to the idea that the government did not prevent contaminated food products from entering the marketplace. It also led to perhaps the earliest example of politicians using food as a political tool, as both Richard Nixon and John F. Kennedy eagerly ate cranberries during the presidential campaign in order to win the support of rural voters in New England, where the cranberry crop had not been tainted.[425]

In 1961, US president Dwight D. Eisenhower also influenced the political climate when he said in his farewell speech,

A vital element in keeping the peace is our military establishment. Our arms must be mighty, ready for instant action, so that no potential aggressor may be tempted to risk his own destruction . . . We have been compelled to create a permanent armaments industry of vast proportions . . . Our toil, resources and livelihood are all involved; so is the very structure of our society. In the councils of Government, we must guard against the acquisition of unwarranted influence, whether sought or unsought, by the military-industrial complex. The potential for the disastrous rise of misplaced power exists and will persist. We must never let the weight of this combination endanger our liberties or democratic processes. We should take nothing for granted. Only an alert and knowledgeable citizenry can compel the proper meshing of the huge industrial and military machinery of defense with our peaceful methods and goals, so that security and liberty may prosper together.[485]

Chemical companies formed a core part of that "military-industrial complex."

A serious drug scare also set the stage for *Silent Spring*. In September 1960, Frances Kelsey, newly employed by the Food and Drug Administration, received an application from the American pharmaceutical company Richardson-Merrell to market the sedative thalidomide in the United States.[486] The drug, originally synthesized in West Germany by Chemie Grünenthal, and licensed for production by drug companies around the world, was used in forty-six countries to treat nausea in pregnant women, as well as more generally for sleep, respiratory ailments, and neuralgia.[467,486] Doctors also prescribed it as a calming agent for children prior to the use of electroencephalograms.[486] Chemie Grünenthal advertised it as the "best drug for pregnant women and nursing mothers."[421]

Studies on animals revealed it to be safe, and Richardson-Merrell requested an expedited review. But Kelsey was disturbed that the drug manifested differently in animals than in people. With every sixty-day deadline for action by the Food and Drug Administration, Kelsey issued a decision of insufficient proof of safety, to the tremendous frustration of Richardson-Merrell, which continually provided further evidence of safety. "She saw her duty in sternly simple terms," wrote a reporter, "and she carried it out, living the while with insinuations that she was a bureaucratic nitpicker, unreasonable—even, she said, stupid."[486]

In the spring of 1961, German researchers were stumped by an inexplicable outbreak of phocomelia, a previously almost unheard-of birth defect that is typically expressed by the lack of an arm and the presence of rudimentary fingers that appear "like the flippers of a seal" below the shoulder.[486] In this outbreak, thousands of babies in Germany and around the world were born without both arms, or without both legs, or without any limbs at all, along with many other defects, and thousands more died. A German pediatrician made the connection to thalidomide on November 3, 1961. It turned out there were even cases in babies born to doctors' wives who had taken thalidomide samples donated by drug companies to their husbands. Richardson-Merrell had distributed 2.5 million pills to 1,200 physicians in the United States after it submitted its application to the Food and Drug Administration, which led to birth defects in the United States. On November 29, the company received the evidence that thalidomide caused birth defects, and it reported this to Kelsey the following day. Richardson-Merrell subsequently withdrew the drug's application, though the company stated that "conclusive proof is lacking for such assumptions."[486] Investigators found that both Richardson-Merrell and Chemie Grünenthal deceived regulators.[421]

The United States largely escaped the fate of thalidomide-induced birth defects, due to the annoying diligence of Kelsey. The larger lesson was that drug companies and medical professionals had aggressively pushed a new chemical onto the marketplace without adequate testing and then criticized a woman who took a more cautious approach. Carson herself saw the parallels; she related in an interview, "It is all of a piece, thalidomide and pesticides. They represent our willingness to rush ahead and use something new without knowing what the results are going to be."[487]

Although disasters primed the public to receive Carson's message, private events conspired to slow her progress.[471] Her mother fell seriously ill, and then died at the end of 1958. At the beginning of 1960, Carson's own health also rapidly declined, but she felt the urgency of finishing the book, and by March the whole text began to take shape. A series of illnesses mounted, and by December she became aware that although her doctor had performed a radical mastectomy the previous spring, the cancer had metastasized. "If one were superstitious," she wrote, "it would be easy to believe in some malevolent influence at work, determined by some means to keep the book from being finished."[471] She continued to write when she was able, noting, "Perhaps more than ever, I am eager to get the book done."[463] Carson described the "two million volt monster" that supplied her radiation as her "only ally . . . but what an awesome and terrible ally, for even while it is killing the cancer I know what it is doing to me."[467]

[**11**]

Silent Spring
(1962–1964)

Pests are simply living organisms distinguished from many other forms of life only by the fact that they have acquired the great displeasure of one of their chief competitors, man . . . Man has annihilated whole armies of his own species, Homo sapiens. Civilizations have come and gone. But it is doubtful if man has ever exterminated, except in local areas, a single one of those competing species he calls pests.—**George C. Decker, prominent scientist and critic of Rachel Carson's** *Silent Spring*, 1962[488]

The chemical war is never won, and all life is caught in its violent crossfire.—**Rachel Carson, 1962[447]**

Titles are the doorway that either obstruct entry or invite readers in. Many potential titles in addition to "Man against the Earth" competed in Carson's mind for the book cover. These included "Man against Nature," "How to Balance Nature," "The Control of Nature," and "Dissent in Favor of Man," the last of which referred to Justice William O. Douglas's dissent of the Supreme Court ruling against Long Island residents who fought aerial spraying of DDT.[425,466] All these would have limited the book's audience, so Carson searched for a better title. Houghton Mifflin's editor in chief, Paul Brooks, had recommended "Silent Spring" as the name for the chapter on birds, and later expanded that suggestion to the book title.[471] Her literary agent, Marie Rodell, recommended to Carson the two lines from the poet John Keats for the front of the book: "The sedge is wither'd from the lake, And no birds sing."[489]

Carson's previous books were beautiful, celebratory, and lyrical works that took the reader along the seacoast and under the waves, but they were not controversial. Controversy, however, lit up *Silent Spring* well before it was published in the fall of 1962. Brooks anticipated its impact, telling Carson when she began writing the book, "One thing you can be sure of: the world is waiting for it."[466] This was due in part to the scale of the problem; by 1962, chemical companies had already registered for the US marketplace about 500 chemical compounds used in 54,000 pesticide formulations, and US insecticide use alone that year amounted to an astonishing 350 million pounds distributed over 90 million acres.[490]

Knowing that the book would face a fierce industrial backlash, Carson vetted all facts with noted experts, and lawyers for Houghton Mifflin assured her that nothing was libelous.[466,471] Carson's team sent prepublication copies to congressional leaders, US government administrators, political organizations, and garden and conservation societies.[467] Battle lines were already drawn by the time the *New Yorker* published a condensed version of the book in June, and *New Yorker* readers submitted a greater volume of letters than any previous piece in the magazine's history had generated.[474] The *New Yorker* boasted a circulation of 430,000 readers, giving *Silent Spring* a strong opening salvo.[466] The *New Yorker*'s editor, William Shawn, who had played a key role in editing the book, noted, "We don't usually think of the *New Yorker* as changing the world, but this is one time it might."[466] The *New York Times* declared "'Silent Spring' Is Now Noisy Summer."[491] The *New Yorker*'s series was one that "few will read without a chill, no matter how hot the weather."[492]

The *New York Times* accurately predicted the effect the book would have, though it was still only in serialized form. "Miss Carson will be accused of alarmism, or lack of objectivity, of showing only the bad side of pesticides while ignoring their benefits. But this, we suspect, is her purpose as well as her method. We do not combat highway carelessness by reciting statistics only of the millions of motorists who return safely to their garages."[492]

On Carson's side stood many prominent figures, including Agnes E. Meyer, who owned the *Washington Post*, and the heads of such women's groups as the League of Women Voters, the National Council of Jewish Women, and the American Association of University Women.[467] She also enjoyed the support of prominent conservation leaders, scientists, and public figures, including Justice William O. Douglas and secretary of the interior Stewart Udall. Douglas declared that *Silent Spring* was "the most revolutionary book since *Uncle Tom's Cabin*."[493] He was quoted on the book's back cover as stating the book to be "the most important chronicle of this century for the human race." The *New York Times* wrote, "If her series helps arouse enough public concern to immunize Government agencies against the blandishments of the hucksters and enforces adequate controls, the author will be as deserving of the Nobel Prize as was the inventor of DDT."[492]

Commercial success of the book seemed guaranteed. It was selected by Book-of-the-Month Club as the October read, excerpts were due to be published in magazines, Consumers' Union contracted for a special edition for its membership, and *CBS Reports* planned a TV episode on the book.[320,471]

Balancing out these positive signs, attacks on Carson and her book rained in well before it was published in September.[467] Velsicol Chemical Company, which produced the insecticides heptachlor and chlordane, threated to sue the *New Yorker* if it continued to print the condensed book.[466,471] The *New Yorker* did not retreat. Velsicol then threatened to sue Houghton Mifflin if it proceeded with publication. Velsicol's letter to Houghton Mifflin stated, "Unfortunately, in addition to the sincere opinions by natural food faddists, Audubon groups and others, members of the chemical industry in this country and in western Europe must deal with sinister influences, whose attacks on the chemical industry have a dual purpose: (1) to create the false impression that all business is grasping and immoral, and (2) to reduce the use of agricultural chemicals in this country and in the countries of western Europe, so that our supply of food will be reduced to east-curtain parity. Many innocent groups are financed and led into attacks on the chemical industry by these sinister parties."[463]

Silent Spring opens with "A Fable for Tomorrow":

> There was once a town in the heart of America where all
> life seemed to live in harmony with its surroundings. The town
> lay in the midst of a checkerboard of prosperous farms, with
> fields of grain and hillsides or orchards where, in spring, white
> clouds of bloom drifted above the green fields. In autumn,
> oak and maple and birch set up a blaze of color that flamed
> and flickered across a backdrop of pines. Then foxes barked
> in the hills and deer silently crossed the fields, half hidden in
> the mists of the fall mornings . . . Then a strange blight crept
> over the area and everything began to change. Some evil spell
> had settled on the community: mysterious maladies swept
> the flocks of chickens; the cattle and sheep sickened and died.
> Everywhere was a shadow of death . . . The few birds seen any-
> where were moribund; they trembled violently and could not
> fly. It was a spring without voices . . . only silence lay over the
> fields and woods and marsh . . . The roadsides, once so attrac-
> tive, were now lined with browned and withered vegetation
> as though swept by fire . . . No witchcraft, no enemy action
> had silenced the rebirth of new life in this stricken world. The
> people had done it themselves.[447]

This destruction wrought by the white clouds drifting over green fields echoed the stark observation of the priest who, in crossing Ireland during the potato famine, "beheld with sorrow one wide waste of putrefying vegetation. In many places the wretched people were seated on the fences of their decaying gardens, wringing their hands and wailing bitterly the destruction that had left them foodless."[21] Carson's fable was also reminiscent of the clouds of chlorine gas blighting the battlefields of Europe during World War I, and could easily invoke everyone's Cold War fears of nuclear fallout. Indeed, she mentioned strontium-90 in the book before referring to pesticides, and used radiation as a metaphor throughout the book.[494] Carson wrote that "the parallel between chemicals and radiation is exact and inescapable."[447] She quoted the 1952 Nobel Peace Prize winner Albert Schweitzer, who said, "Man can hardly even recognize the devils of his own creation."[447] "The question is," posed Carson, "whether any civilization can wage relentless war on life without destroying itself, and without losing the right to be called civilized."[447]

Monsanto Corporation responded with a widely distributed parody entitled "The Desolate Year," which described a pesticide-free world riddled with disease and hunger.[495] "The bugs were everywhere. Unseen. Unheard. Unbelievably universal . . . Beneath the ground, beneath the waters, on and in limbs and twigs and stalks, under rocks, inside trees and animals and other insects—and, yes, inside man." Due to the lack of pesticides, "the garrote of Nature rampant began to tighten." The result was that "genus by genus, species by species, sub-species by innumerable sub-species, the insects emerged. Creeping and flying and crawling into the open, beginning in the southern tier of states and progressing northward." People, "infected by the first onslaught of the host mosquitoes, suffered the fiendish torture of chills and fever and the hellish pain of the world's greatest scourge." Malaria was by no means the only agent of man's suffering. "Then the really notorious villain, Ireland's awful late blight, took over, and the firm brown 'spuds' were gone, turned into black slime." A repeat of the Irish Potato Famine due to a lack of pesticides led to starving people once again reduced to eating insects. Termites felled buildings and devoured libraries. "Yellow fever hung like a spectre" over the southern United States. "Rats and mice multiplied prodigiously," a disaster that would lead to outbreaks of typhus and bubonic plague.

Similarly, the *American Agriculturist* featured a boy and his grandfather eating acorns in the forest. The grandfather explained that a book had come out against the use of chemicals in farming. "So now we live naturally. Your mother died naturally from malaria that mosquitoes gave her; your Dad passed away naturally in that terrible famine when the grasshoppers ate up everything; now we are starving naturally, because the blight killed those potatoes we planted last spring."[471]

A host of chemical companies, including Monsanto, DuPont, Dow, Shell Chemical, Goodrich-Gulf, Allied Chemical, and W. R. Grace, collaborated through trade organizations in their criticisms of the book and its author, and some, such as Velsicol and American Cyanamid, had their own representatives mount attacks.[466] *Silent Spring* threatened to undermine the prestige of these companies, carefully cultivated through advertising campaigns such as DuPont's "Better Living through Chemistry." Industry also worried that the book would lead to unwanted regulations. *Chemical and Engineering News* quoted the director of the New Jersey Department of Agriculture: "In any large scale pest control program in this area, we are immediately confronted with the objection of a vociferous, misinformed group of nature-balancing, organic-gardening, bird-loving, unreasonable citizenry."[471] Another magazine concluded, "Her book is more poisonous than the pesticides she condemns."[471] The irony of the industry campaign was revealed by one critic of Carson, who wrote, "They scold emotion emotionally."[496]

Carson wrote, "DDT is now so universally used that in most minds the product takes on the harmless aspect of the familiar."[447] Such indiscriminate use of pesticides by government agencies and private parties posed serious ethical issues beyond the effects on nature, and this theme was central to *Silent Spring*. Ordinary people were exposed to pesticides without their knowledge or consent. By 1950, Americans already averaged over 5 parts per million DDT in their body fat, and women's breast milk was also contaminated.[320] The concentration of DDT and its metabolites in the body fat of the average American adult rose to 12 parts per million by the early 1960s.[490] Carson argued that people had a right to know and a role to decide about pesticide use, both of which were denied them. She wrote that we live in "an era dominated by industry, in which the right to make a dollar at whatever cost is seldom challenged. When the public protests, confronted with some obvious evidence of damaging results of pesticide applications, it is fed little tranquilizing pills of half truth . . . It is the public that is being asked to assume the risks that the insect controllers calculate."[447]

This risk was extended even into their own homes by their own hands, given the universal availability of pesticides for domestic use, placing the typical household "in little better position than the guests of the Borgias."[447] One industry executive essentially agreed, stating, "The industry deserves a black eye for not educating pesticide users on the proper use of powerful chemicals. One of the big problems we've always faced is overestimating the intelligence of users."[466] Carson wrote, "If the Bill of Rights contains no guarantee that a citizen shall be secure against lethal poisons distributed either by private individuals or by public officials, it is surely only because our forefathers, despite their considerable wisdom and foresight, could conceive of no such problem."[447] She also extended such ethical concerns to animals and wrote, "By acquiescing in an act that can cause such suffering to a living creature, who among us is not diminished as a human being?"[447]

Silent Spring ends with the paragraph:

The "control of nature" is a phrase conceived in arrogance, born of the Neanderthal age of biology and philosophy, when it was supposed that nature exists for the convenience of man. The concepts and practices of applied entomology for the most part date from that Stone Age of science. It is our alarming misfortune that so primitive a science has armed itself with the most modern and terrible weapons, and that in turning them against the insects it has also turned them against the earth.[447]

Applied entomologists were thus directly attacked as ignorant and immoral. They and their allies responded in kind. One prominent entomologist wrote that "*Silent Spring* poses leading questions, on which neither the author nor the average reader is qualified to make decisions. I regard it as science fiction, to be read in the same way that the TV show *Twilight Zone* is to be watched."[467] An industrial trade journal commented, "For the insecticide industry, this book could turn out to be a serious and costly body blow—even though it did land below the belt."[467]

A prominent scientist, who led the Food Protection Committee of the National Academy of Sciences–National Research Council, predicted that *Silent Spring* would appeal to "the organic gardeners, the antifluoride leaguers, the worshipers of 'natural foods,' those who cling to the philosophy of a vital principle, and pseudo-scientists and faddists."[497] He advised that, "in view of her scientific qualifications in contrast to those of our distinguished scientific leaders and statesmen, this book should be ignored . . . It is doubtful that many readers can bear to wade through its high-pitched sequences of anxieties."[497] He warned that the attitude expressed in the book "means the end of all human progress, reversion to a passive social state devoid of technology, scientific medicine, agriculture, sanitation, or education. It means disease, epidemics, starvation, misery, and suffering incomparable and intolerable to modern man."[497]

The head of the nutrition program at the Harvard School of Public Health stated:

> Miss Carson writes with passion and with beauty, but with very little scientific detachment. Dispassionate scientific evidence and passionate propaganda are two buckets of water that simply can't be carried on one person's shoulders. The bucket that springs a leak in Miss Carson's case is the scientific evidence . . . Unfortunately, Miss Carson implicitly accuses the scientific community of disparaging human values. In doing so, she abandons scientific proof and truth and combats them with exaggeration and unscientific deductive reasoning based

on axioms of her own making . . . Miss Carson is a literary luminary—and one of splendid accomplishment. That is no mean feat. Miss Carson could have written a book which would have helped bridge the gap between science and the public, instead of one which widens the gulf.[498]

The president of Montrose Chemical Corporation, which was the largest DDT producer in the United States, accused Carson of being "a fanatic defender of the cult of the balance of nature."[491] Many others jumped on this theme, such as a government official who said, "At one time this country was supporting about 1,000,000 Indians and some wildlife, and things were in balance."[499] A trade organization called the National Agricultural Chemicals Association funded a $250,000 public relations campaign attacking Carson.[474] A former secretary of agriculture asked "why a spinster with no children was so concerned about genetics?" His answer was that Carson was "probably a Communist."[467]

The most visible critic was Robert White-Stevens of American Cyanamid. He wrote, "We often find that a false statement takes off like an ICBM [intercontinental ballistic missile] and explodes in TV, radio, newspapers, magazines, and even in books, before a measured estimate of the facts in the matter can be assessed and presented objectively."[500] White-Stevens admired Carson's writing, but not her intent:

Miss Rachel Carson . . . is a writer on biological subjects with an extraordinary, vivid touch and elegance of expression. She paints a nostalgic picture of Elysian life in an imaginary American village of former years, where all was in harmonious balance with Nature and happiness and contentment reigned interminably, until sickness, death, and corruption was spread over the face of the landscape in the form of insecticides and other agricultural chemicals. But the picture she paints is illusory, and she as a biologist must know that the rural Utopia she describes was rudely punctuated by a longevity among its residents of perhaps thirty-five years, by an infant mortality of upwards of twenty children dead by the age of five of every 100 born, by mothers dead in their twenties from childbed fever and tuberculosis, by frequent famines crushing the isolated peoples through long dark, frozen winters following the failure of a basic crop the previous summer, by vermin and filth infesting their homes, their stored foods and their bodies, both inside and out.[500]

Criticism did not limit itself to the pen of entomologists and industrialists. Many media outlets joined the fray. The *Economist* reduced Carson's "angry, shrill tract" to "propaganda written in white-hot anger with words tumbling and stumbling all over the page."[474] *Time* magazine's science journalist charged that in her "emotional and inaccurate outburst," "Miss Carson has taken up her pen in alarm and anger, putting literary skill second to the task of frightening and arousing her readers," leading to a position that was "unfair, one-sided, and hysterically overemphatic."[501] Carson wrote, "These insecticides are not selective poisons; they do not single out the one species of which we desire to be rid. Each of them is used for the simple reason that it is a deadly poison. It therefore poisons all life with which it comes in contact."[447] The *Time* journalist responded, "Any housewife who has sprayed flies with a bug bomb and managed to survive without poisoning should spot at least part of the error in that statement."[501] This journalist and others drew upon the experience of convicts who were fed DDT by the US Public Health Service and were found to be as healthy as a control group of convicts fed the same diet minus DDT.[502]

Natural History magazine provided a mixed review, though the author defended Carson's advocacy: "She has been accused of being 'one-sided,' as though this were a fault. I have never heard` St. Paul criticized for not giving Satan his due, though he is obviously a devilishly engaging fellow."[503] The secretary of the interior, Stewart Udall, defended Carson along these same lines: "*Silent Spring* was called a one-sided book. And so it was. She did not pause to state the case for the use of poisons on pests, for her antagonists were riding roughshod over the landscape. They had not bothered to state the case for nature. The engines of industry were in action; the benefits of pest control were known—and the case for caution needed dramatic statement if alternatives to misuse were to be pursued."[504]

Edwin Diamond, the science editor of *Newsweek*, initially was contracted by Houghton Mifflin to cowrite the book with Carson, but Carson decided the collaboration would only hurt the book and so removed him from the project early on.[466] Diamond ended up writing one of the most scathing reviews.[487] "Thanks to a woman named Rachel Carson," he wrote, "a big fuss has been stirred up to scare the American public out of its wits." Noting that Carson was not married, Diamond asked, "What, finally, is *Silent Spring's* game? . . . this is an era of stereotyped thinking, scattershot charges, shrill voices and double standards of behavior." Carson's tactics, Diamond wrote, were reminiscent of those employed by Senator Joseph McCarthy in his Great Communist Hunt. "The record shows that the nation, once down from its McCarthyite orbit, was able to deal with subversion without dismantling its noble mansion of constitutional law and civil rights. Similarly, I think the pesticide 'problem' can be handled without going back to a dark age of plague and epidemic."

The controversy fell in part along gender lines, with men doing the criticizing and employing sexist stereotypes, and women (and men) providing critical support, although also often in a gender-biased manner. Carson was compared to Carrie Nation, the hatchet-wielding activist against alcohol.[505] The journal *Archives of Internal Medicine* published an editorial stating, "*Silent Spring*, which I read word for word with some trauma, kept reminding me of trying to win an argument with a woman. It can not be done."[496] Although this author found the book, as science, to be "so much hogwash," he also expressed a view that some positive changes were likely to result from it. "I found in it much cantankerous enthusiasm and much concern for the state of man. I got through it, realizing that it will not advance the cause of science, of learning, or of Rachel Carson. On the other hand, if enough people read it and begin to think about the implications of some of the things of which it gives only a biased view, not only the country, but even scientists working on the problems the book assails, must surely gain."[496]

Gender stereotypes also colored public perceptions. The *New York Times* wrote, "Gentle, soft-voiced Rachel Louise Carson appears an unlikely candidate for the role of avenging angel. She is shy and very feminine, and refuses to be drawn into retaliatory comments against those who have attacked her book . . . Miss Carson is a small-boned woman with gradually graying, dark-brown hair. Her eyes are grayish brown and her complexion is pale."[506] Another friendly reviewer wrote, "Miss Carson is a quiet-spoken spinster, now 55 years old, who lives in suburban Silver Springs, Md., near Washington. She is renowned for her prize-winning 1951 work on marine biology, 'The Sea Around Us,' dealing with the flora and fauna of the sea. In her latest books she comes ashore, with a vengeance."[499] But Carson dismissed any role for gender in the point of her book; her interest was not in "things done by women or men, but in things done by people."[467]

In perhaps the only serious criticism by a woman, Virginia Kraft wrote in *Sports Illustrated*, "Wildlife populations all over the nation are bigger and healthier than ever, not in spite of pesticides, but in many cases because of them . . . The prosperity in the wildlife of today is a direct result of man's—particularly American man's—increased ability to control his own environment . . . The single most effective tool in bringing about these improvements has been chemical pesticides."[507] Kraft wrote that *Silent Spring* made Nevil Shute's nuclear apocalypse novel *On the Beach* "seem almost euphoric by comparison."[507] She continued, "The wise and discreet use of chemical pesticides is assuring us not of a silent spring, but of seasons filled with all the rich, new sounds of animal and human prosperity."[507]

In summing up the criticism originating from so many quarters, Paul Brooks compared the reaction to that faced by Charles Darwin a century earlier with the publication of *On the Origin of Species*; not since then "had a single book been more bitterly attacked by those who felt their interests threatened."[471] Even some of the language used to criticize Darwin rang familiar in 1962. Louis Agassiz wrote that Darwin's transmutation theory is "a scientific mistake, untrue in its facts, unscientific in its method, and mischievous in its tendency."[508] Comparing *Silent Spring* to Thomas Paine's *Rights of Man*,[509] a critic for the *New York Times* noted, "From the panic of the chemical industries and the soothing officialese of Government bulletins, you would think that Rachel Carson had advocated a return to the wooden plow."[510]

All this controversy stimulated sales of the book and thereby worked against the critics. Within two months of publication, the book had sold 100,000 copies, topped the bestseller list, and featured in headline news; indeed, the book caught the attention of President Kennedy even before it was published.[467] It was translated into twenty-two languages, and half a million hardcover books were sold before Houghton Mifflin released the paperback edition.[467,474] Sales occurred in pulses with each new event, whether it was the airing of *CBS Reports*, or press coverage of government hearings. A Houghton Mifflin executive noted this unprecedented situation for one of its titles, and stated, "The only parallel I can think of is a book as evil as Rachel's is benign. *Mein Kampf* used to have a new wave of sales whenever Hitler invaded a new country. People were trying to see who would suffer next."[466] Sensitive to the charges of a communist conspiracy, Carson and her agent decided not to sell literary rights to communist countries "because the book is too easily twisted to anti-U.S. propaganda."[466]

Too ill to do much, Carson agreed to only a handful of events and interviews.[467] She wrote to her doctor, "I still believe in the old Churchillian determination to fight each battle as it comes, and I think a determination to win may well postpone the final battle."[463] One opportunity for battle that she found irresistible was an interview with *CBS Reports*, which, due to its popularity and prime-time slot, would gain a wide audience for her views. Since Carson was limited in mobility because of her cancer, the show's host and his crew conducted the interview in her home. CBS constructed the show through eight months of filming.[425] Opposing Carson, *CBS Reports* featured White-Stevens as the representative of the chemical companies. Several government leaders were also featured: the surgeon general, the commissioner of the Food and Drug Administration, and the secretary of agriculture. The controversial nature of the content led three major corporations to withdraw their advertisements.

The show aired on April 3, 1963, and vindicated Carson as the calm adult in the room, not as the hysterical communist spinster. Carson's voice opened the program: "Can anyone believe it is possible to lay down such a barrage of poisons on the surface of the earth without making it unfit for all life? They should not be called 'insecticides' but 'biocides.'"[511] White-Stevens responded, "The major claims in Miss Rachel Carson's book *Silent Spring* are gross distortions of the actual facts, completely unsupported by scientific experimental evidence and general practical experience in the field . . . If man were to faithfully follow the teachings of Miss Carson, we would return to the Dark Ages, and the insects and diseases and vermin would once again inherit the Earth." White-Stevens continued, "Miss Carson maintains that the balance of nature is a major force in the survival of man whereas the modern chemist, the modern biologist, the modern scientist believes that man is steadily controlling nature." Carson retorted, "Now to these people apparently the balance of nature was something that was repealed as soon as man came on the scene. Well you might just as well assume that you could repeal the law of gravity . . . Man is part of nature and his war against nature is inevitably a war against himself." Carson ended the show by saying, "I think we're challenged as mankind has never been challenged before, to prove our maturity and our mastery, not of nature, but of ourselves."

The show drew an audience of 10 million viewers.[474] The day after the show aired, Senator Abraham Ribicoff of Connecticut declared that his Government Operations Subcommittee would hold hearings on the dangers of pesticides.[467] The hearings began on May 15, and Carson, the most anticipated witness called to testify, appeared on June 4. Ribicoff said, "Miss Carson, you are the lady who started all this."[474] His comment was reminiscent of Abraham Lincoln's question to Harriet Beecher Stowe, the author of *Uncle Tom's Cabin*. According to the family lore of Stowe's descendants, Lincoln said when he met her, "Is this the little woman who made this great war?"[512] In her testimony, Carson stated, "Since our problems of pest control are numerous and varied, we must search, not for one superweapon that will solve all our problems, but for a great diversity of armaments, each precisely adjusted to its task."[463] Carson advocated restrictions on aerial spraying and persistent pesticides, the creation of a government body responsible for testing and control, and the incorporation of public input so that people would be secure from poisoning in their homes. Carson added that scientists had just discovered pesticides in remote regions of the globe far from their point of application. The problem of pesticide applicators indiscriminately spraying without the knowledge or consent of the public therefore extended to communities that might not even have heard of the countries where the pesticides originated. A consultant for the Shell Chemical Company countered her testimony and stated, "These peddlers of fear are going to feast on the famine of the world."[463]

The same day that the Ribicoff hearings began, the President's Science Advisory Committee reported about the benefits and risks of pesticides. The report noted, "Efficient agricultural production, protection of health, and elimination of nuisances are now required and expected by modern man."[490] An example of many positive developments was that "sweet corn, potatoes, cabbage, apples, and tomatoes are all available unmarred, and the American housewife is accustomed to blemish-free products."[490] But the report also noted the problem of resistance, and stated, "Until the publication of 'Silent Spring' by Rachel Carson, people were generally unaware of the toxicity of pesticides."[490] Carson felt relieved by the content of the report, which concluded, "Elimination of the use of persistent toxic pesticides should be the goal."[490] Two days later, she testified before the Senate Committee on Commerce, which was debating pesticide regulations.[471] Carson's recommendation was the creation of a cabinet-level agency for environmental regulations that would be independent of chemical industry influence.

The Science Advisory report turned the tide on public discourse in her favor, with headlines reading "Rachel Carson Stands Vindicated," and some critics in the media admitting that perhaps she was right.[463] *CBS Reports* broadcast a follow-up program entitled "The Verdict on the Silent Spring of Rachel Carson."[466] Lord Shackleton, the son of the Antarctic explorer, wrote the introduction to the British edition of *Silent Spring*. He stated in the House of Lords that Polynesian cannibals would no longer eat Americans "because their fat is contaminated with chlorinated hydrocarbons."[471] Data on DDT concentrations showed, said Lord Shackleton, that "we [British] are rather more edible than Americans." Britain's response to the book was to expand pesticide regulations.

Newsworthy events also vindicated Carson. At the end of 1963, 5 million fish in the Mississippi River died in the midst of convulsions and hemorrhaging.[463] Scientists traced the mortality to releases of the pesticide endrin by a pesticide plant operated by Velsicol, the chemical company that had first developed endrin and that had threatened to sue over the release of *Silent Spring*. Carson wrote that wildlife mortality due to pesticide exposure did not end with the publication of her book. "The problem of pesticides is not merely the dream of an avaricious author, out to pile up royalties by frightening the public," she wrote, "it is very much with us, here and now."[513] That DDT progressed from mankind's savior in World War II to one of humanity's greatest toxic burdens thirty years later demonstrates the wild swings in public perception and the chaotic regulatory environment surrounding pesticides.

A series of awards in 1963 marked the end of Carson's career.[471] At the beginning of the year, the Animal Welfare Institute gave her the Albert Schweitzer medal. Many other awards followed, along with her election to the American Academy of Arts and Letters. Academy membership was restricted to fifty artists, musicians, and writers; as a new member, Carson joined the company of just three other women, and no other writers of nonfiction. The academy citation summarized Carson's contribution: "A scientist in the grand literary style of Galileo and Buffon, she has used her scientific knowledge and moral feeling to deepen our consciousness of living nature and to alert us to the calamitous possibility that our short-sighted technological conquests might destroy the very sources of our being."[471] In a speech to the Women's National Book Association, Carson explained why she felt compelled to write the book. "If I had not written the book," she said, "I am sure the ideas would have found another outlet. But knowing the facts as I did, I could not rest until I had brought them to public attention."[463]

Carson dedicated *Silent Spring* to Albert Schweitzer. Schweitzer wrote to a beekeeper, "I am aware of some of the tragic repercussions of the chemical fight against insects taking place in France and elsewhere, and I deplore them. Modern man no longer knows how to foresee and forestall. He will end by destroying the earth from which he and other living creatures draw their food. Poor bees, poor birds, poor men."[514] From this statement sprang Carson's dedication to Schweitzer: "Man has lost the capacity to foresee and to forestall. He will end by destroying the earth."[447] Following the publication of *Silent Spring*, Schweitzer wrote a letter of thanks to Carson, and included his photograph, which became Carson's most cherished possession.[493]

Knowing that she would soon die, Carson requested that a passage from *The Edge of the Sea* be read at her funeral:[471] "What is the meaning of so tiny a being as the transparent wisp of protoplasm that is a sea lace, existing for some reason inscrutable to us—a reason that demands its presence by the trillion amid the rocks and weeds of the shore? The meaning haunts and ever eludes us, and in its very pursuit we approach the ultimate mystery of Life itself."[475]

Carson died on April 14, 1964, at the age of fifty-six.[471] Her pallbearers included Secretary of the Interior Udall and Senator Ribicoff. Ribicoff paid tribute on the floor of the Senate to "this gentle lady who aroused people everywhere to be concerned with one of the most significant problems of mid-twentieth century life—man's contamination of his environment."[463] Paul Brooks and his wife raised her orphaned grandnephew, and Brooks published a book devoted to her writing.[466,471] In it, he wrote of Carson's strength in completing *Silent Spring* as her health failed. Brooks wrote, "She managed to make this book about death a celebration of life."[471]

"Rachel Carson is dead," wrote E. B. White, "but the sea is still around us, the edge of the sea still supports life in almost unbelievable variety, and the manufacturers of pesticides are enjoying their usual spring up-surge in sales."[515] A year after her death, the chemical company Velsicol stated, "In case you haven't noticed, trees leafed, birds sang, squirrels reconnoitered, fish leaped—1965 was a normal spring, not the 'silent type' of the late Miss Carson's nightmares."[463]

With the publication of *Silent Spring*, Rachel Carson joined a long list of accomplished American authors published by Houghton Mifflin.[466] These included Henry Wadsworth Longfellow, Oliver Wendell Holmes, Ralph Waldo Emerson, Harriet Beecher Stowe, Nathaniel Hawthorne, Henry David Thoreau, and Mark Twain. The publishing house was also the American publisher for such English greats as Alfred, Lord Tennyson; Charles Dickens; and Winston Churchill. Like the works of these great authors, *Silent Spring* became a classic, a book that marked a shift in public awareness throughout the world about the environment, about governmental accountability, about democracy, and about animal and human rights. "Who has decided," Carson asked, "who has the *right* to decide—for the countless legions of people who were not consulted that the supreme value is a world without insects, even though it be also a sterile world ungraced by the curving wing of a bird in flight? The decision is that of the authoritarian temporarily entrusted with power; he has made it during a moment of inattention by millions to whom beauty and the ordered world of nature still have a meaning that is deep and imperative."[447]

Perhaps Carson's message was best foreshadowed by President Eisenhower, who said in his farewell speech to the nation, "As we peer into society's future, we—you and I, and our Government—

must avoid the impulse to live only for today, plundering, for our own ease and convenience, the precious resources of tomorrow. We cannot mortgage the material assets of our grandchildren without risking the loss also of their political and spiritual heritage. We want democracy to survive for all generations to come, not to become the insolvent phantom of tomorrow."[485]

Wonder and Humility

(1962–THE FUTURE)

Within the past 100 years, man has emerged from a feeble creature, virtually at the mercy of Nature and his environment, to become the only being which can penetrate every corner of the planet, communicate instantly to anywhere on earth, produce all the food, fiber, and shelter he needs, wherever he may need it, change the topography of his lands, the sea and the universe and prepare his voyage through the very arch of heaven into space itself. This is the stuff that science is made of, and man has learned to use it.—**Robert H. White-Stevens**, representative of the chemical industry, 1962[500]

Following World War II, chemists continued to develop products that would make the world a better place by preventing famine and disease; they also continued to design potent chemical weapons for the next war. War had become so frequent and horrific that it seemed foolish to expect it to retreat for long or to again be caught unprepared for a dictator's onslaught.

Together, these chemicals designed for both the best and worst purposes of humanity proved to be an environmental disaster and compelled the writing of Rachel Carson. But the storm surrounding the accelerated development and deployment of pesticides did not end with Carson's death in 1964.

Rachel Carson wrote to a friend, as she began writing *Silent Spring*, "There would be no peace for me if I kept silent."[516] As she completed the book, she wrote in a letter, "I could never again listen happily to a thrush song if I had not done all I could."[471] The same compulsive drive likely plagued, motivated, and enabled the greatness of all of the scientists in this book. Ronald Ross strained his vision peering through a microscope for years before he made his monumental discovery of pigmented cells in the stomach wall of an *Anopheles* mosquito, literally as his microscope rusted from the drips of his sweat. Fritz Haber and Walther Nernst risked fiery death in their high-pressure and high-heat competition to fix nitrogen from the atmosphere. Gerhard Schrader exposed himself to the most toxic compounds ever synthesized while he manipulated nature to make organophosphate insecticides.

Many traumatic events involving chemical exposures occurred following the publication of *Silent Spring*. Scientists discovered DDT in women's breast milk at concentrations five times greater than the maximum allowable level in cow's milk.[463] The Environmental Defense Fund, created by an informal group of Long Island environmentalists, scientists, and a lawyer in the wake of Carson's death, placed an advertisement in the *New York Times* that asked, "Is Mother's Milk Fit for Human Consumption?"[320] Similarly, the Ecology Center in Berkeley produced a poster of a nude pregnant woman with a label on her breast that read, "Caution, keep out of the reach of children."[463]

Elsewhere in the world, in the summer of 1967, 700 people were hospitalized and 24 died in Qatar after eating bread made from American flour that had been shipped in cloth sacks on a deck below leaking pails of endrin.[463] That same year, 17 children died and 600 others sickened in Mexico after eating pastries prepared using sugar that had been stored adjacent to parathion, and a similar accident with parathion-contaminated bread killed 80 and sickened

600 in Colombia. Similar tragedies with a variety of pesticides occurred throughout the world, and became more common and caused more fatalities as the variety of new pesticides in the marketplace proliferated and the volumes employed jumped. Indeed, in 1966, four years after the publication of *Silent Spring*, the mass of synthetic pesticides sold by US companies surpassed 840 million pounds, while inorganic pesticides contributed another 450 million pounds to the US market.[463]

Market pressures for new chemicals used in larger quantities resulted from pests evolving resistance. The evolution of resistance to DDT on the part of *Anopheles* mosquitoes led the World Health Organization to suspend the Global Malaria Eradication Programme in 1969.[517] The chemical industry benefited from the inherent obsolescence of pesticides, in the same way that the armaments industry benefited from the arms race inducing obsolescence of weapons.

Pesticides also killed people through disruptions of the "balance of nature," but in rather unpredictable ways. One striking example occurred in Bolivia in 1963, where more than 300 people perished from an outbreak of Bolivian hemorrhagic fever.[463] The deadly virus was transmitted by a local rodent species. The rodent population had been held in check by cats, until the cat population crashed due to DDT exposure from an antimalaria campaign.

Operation Ranch Hand proved to be the largest and most enduring pesticide disaster coming on the heels of *Silent Spring*. It began the same year that *Silent Spring* went to press and continued until 1971.[518] In the operation, the United States sprayed 73 million liters of herbicides and defoliants, such as Agent Orange, over the rainforest and mangrove forest canopies of Vietnam, Laos, and Cambodia.[519,520] Crop lands made up 10 percent of the 26,000 square kilometers sprayed. The objective was to deprive Viet Cong forces of food while also removing the forest canopy that obscured their movements. Smokey the Bear became the unofficial mascot of the operation, with a modified slogan that read, "Only *you* can prevent a forest."[521] Aircraft that dropped smoke grenades or flares to mark forest to be defoliated or enemy positions were nicknamed "Smokey

Bears." In the related Sherwood Forest (1965) and Pink Rose (1966) operations, US forces defoliated the rainforest to dry out trees and then dropped diesel fuel and incendiary devices to initiate fire as a weapon against the Viet Cong. Because the spray aircraft flew slowly, their missions were typically preceded by aerial bombardments with bombs, ordnance, and napalm.[520]

The operation reflected an attitude discussed by Carson in *Silent Spring*, and one noticed by many critics of the military's uses and abuses of new technologies. Carson wrote, "The chemical weed killers are a bright new toy. They work in a spectacular way; they give a giddy sense of power over nature to those who wield them, and as for the long-range and less obvious effects—these are easily brushed aside as the baseless imaginings of pessimists."[447]

The use of chemical defoliants in war was a possibility that emerged from the discovery of plant hormones that regulate growth.[520] In 1941, Ezra Kraus, a botanist on the faculty of the University of Chicago, proposed that synthetic versions of these plant hormones could be used in high doses as herbicides. Kraus and the US Agricultural Research Center screened chemicals for this purpose, including the compound 2,4-D, which had recently been shown to stimulate plant growth. Just days after the US entry into World War II, Kraus proposed that the United States should synthesize herbicides that would enable a "simple means of destruction of rice crops, the staple food supply of the Japanese."[520] These chemicals could also be sprayed over forests to kill trees and thereby "reveal concealed military depots." Kraus's work at Chicago occurred a stone's throw away from the field where Enrico Fermi built the first nuclear reactor.

The Chemical Warfare Service expanded the screening program for herbicides useful in combat as crop destroyers or defoliants. By the end of the war, the Chemical Warfare Service had tested about one thousand chemicals and found 2,4-D and 2,4,5-T to be the most effective. The war ended, and rather than being used for combat, these herbicides were commercialized as weedkillers. The chemical 2,4,5-T did see use in combat in the early 1950s in Malaya, where British forces employed the herbicide to destroy crops and defoliate trees. Thus, the stage was set for the US use of Agent Orange in the Vietnam War.

Following the publication of *Silent Spring*, the President's Science Advisory Committee recommended that pesticides be tested for their potential to cause cancer, birth defects, or genetic defects.[522] Therefore, in 1963 the National Cancer Institute funded the Bionetics Research Laboratories to test the toxicity of selected pesticides. One of the tested pesticides was 2,4,5-T, by then a popular weedkiller and, along with 2,4-D, the active ingredient in Agent Orange. In 1966, Bionetics alerted the National Cancer Institute that 2,4,5-T induced birth defects in mice. Despite widespread use of the herbicide in the United States and in Vietnam, the government did not release this information to the public. Another report from Bionetics to the National Cancer Institute followed in 1968. Still, the information remained confined to a small number of scientists, government regulators, and members of the pesticide industry.

Nevertheless, some scientists expressed alarm at the use of defoliants in the war, and in 1967 the American Association for the Advancement of Science petitioned the US secretary of defense Robert McNamara to authorize a study of effects of defoliant use in Vietnam.[522] The government response stated, "Qualified scientists, both inside and outside our Government, and in the governments of other nations, have judged that seriously adverse consequences will not occur. Unless we had confidence in these judgements, we would not continue to employ these materials."[522] Finally, in the fall of 1969, an employee with a group funded by the consumer advocate Ralph Nader chanced upon a copy of the Bionetics report and relayed it to the Harvard biologist Matthew Meselson.

Meselson had previously challenged the position of the United States on chemical and biological weapons.[522] He had also read newspaper accounts of sharp increases in birth defects in areas of Vietnam sprayed with Agent Orange. Meselson visited the physicist Lee DuBridge, President Nixon's science advisor, on October 29, 1969, to relay his concerns. With Meselson still in his office, DuBridge called the deputy secretary of defense and cofounder of the Hewlett-

Packard Company, David Packard, and they decided to restrict the use of 2,4,5-T.[520] DuBridge issued a press release that very day stating that the Department of Defense "will restrict the use of 2,4,5-T to areas remote from population," while the Department of Agriculture "will cancel registrations of 2,4,5-T for food crops effective January 1" and the Departments of Agriculture and Interior "will stop using 2,4,5-T in their own programs."[520]

DuBridge called Meselson a few days later to relate that Dow Chemical Company concluded that dioxin was to blame, not the herbicide 2,4,5-T.[520] Dioxin was present in 2,4,5-T due to faults in the production process, so the Departments of Agriculture and Defense stated that with improved production techniques they could continue to use the herbicide. Dow Chemical then, in early 1970, conducted research that demonstrated that the improved 2,4,5-T did not cause birth defects.[522] The US Food and Drug Administration and the National Institutes of Health tried to replicate the Dow results, but they found that 2,4,5-T, like thalidomide, strongly induced birth defects, though the effects were even stronger when dioxin was present. The confirmation of the original 1966 Bionetics report took only six weeks to achieve after years of inaction and obfuscation of the findings. Hearings in the Senate followed. However, a President's Science Advisory Committee report stated, "The lack of accurate epidemiological data on the incidence and kinds of birth defects in the Vietnamese population before or since the military use of defoliants precludes any estimate as to whether an increase in birth defects has occurred."[522]

Meanwhile, Dow Chemical and Hercules Corporation appealed the modest restrictions on the use of 2,4,5-T, which led to another advisory committee of scientists recommended by the National Academy of Sciences.[522] This committee included scientists employed by manufacturers of 2,4,5-T. The committee recommended to the newly created Environmental Protection Agency the continued use of 2,4,5-T synthesized to prevent dioxin contamination.

After years of petitioning by concerned scientists, the American Association for the Advancement of Science funded an investigation in 1970.[522] The four-member scientific team, known as the Herbicide Assessment Commission, was organized by Meselson, directed by Arthur Westing, and also included John Constable and Robert Cook. (This book's author had the good fortune to work for Westing in 1987.)

The team traveled to Vietnam and found that contrary to declarations by the Department of Defense, spraying had destroyed crops in densely populated areas, severely damaged half of Vietnam's mangrove forests and one-fifth of its hardwood forests, and appeared to induce stillbirths and birth defects among children exposed in the womb.[519,522,523] The use of herbicidal weapons, wrote Westing, "results in extended human suffering to noncombatants all out of proportion to any immediate military benefits that could be claimed for them."[524]

The team found moral support from thousands of scientists, including seventeen Nobel Laureates, who petitioned the United States to renounce herbicidal weapons.[524] The accumulation of evidence and the team's report led to the suspension of registrations for 2,4,5-T in US agriculture in 1970, along with the discontinuation of Agent Orange operations in Vietnam.[520] However, the use of herbicides to destroy crops continued until January 7, 1971, despite the conclusion of an interagency review in 1968 that read, "The main impact of crop destruction, however, falls upon the civilian population . . . An estimated 90% of the crops destroyed in 1967 were grown, not by VC/NVA [Viet Cong/North Vietnamese Army] military personnel, but by civilians living there."[520]

A follow-up study by the National Academy of Sciences, which Meselson contributed to, reported in 1974 that the defoliation program caused illnesses and deaths among children, damaged mangrove forests to such an extent that they would take a century to recover, increased the prevalence of malaria-infected mosquitoes, and devastated food supplies leading to population displacements.[522] President Gerald Ford signed Executive Order 11850 in 1975: "The United States renounces, as a matter of national policy, first use of herbicides in war except use, under regulations applicable to their domestic use, for control of vegetation within U.S. bases and installations or around their immediate defensive perimeters."[520]

These events were facilitated by the 1969 renunciation of first use of chemical weapons by President Richard Nixon on behalf of the United States, along with a complete ban on the use of biological weapons.[303] Meselson influenced both decisions, as he had access to the president through his former Harvard colleague, Henry Kissinger, who held key posts in both the Nixon and Ford administrations.[522] Nixon called on the Senate to ratify the Geneva Protocol of 1925, stating that "mankind already carries in its own hands too many of the seeds of its own destruction."[303] The Senate followed suit, and also ratified the Biological Weapons Convention, which was the first to ban an entire family of weapons in the aftermath of World War II. President Gerald Ford signed both conventions in 1975, the same year in which he banned first use of herbicidal weapons.

Other chemical events intruded upon the day-to-day lives of people throughout the world and continued the build-up of skepticism in government and industry that Carson launched. One of these events occurred on March 13, 1968, when a test of VX nerve gas at the US Army's Dugway Proving Ground in Utah resulted in the death of six thousand sheep grazing as far away as forty-five miles from the site.[522] Over the next fourteen months, and in an atmosphere of incontrovertible facts, the army's position migrated from outright denial of nerve gas testing to paying damages to ranchers.

In 1969, the US Army prepared to ship 27,000 tons of chemical weapons from the Rocky Mountain Arsenal to the Atlantic Ocean aboard 800 railroad cars, for marine disposal.[522] The shipment was to include 12,000 tons of sarin-filled bombs and 2,600 tons of leaking sarin-filled rockets encased in concrete and steel. New York representative Max McCarthy, concerned about the prudence of shipping weapons of mass destruction across an unreliable rail network, raised a public alarm. A National Academy of Sciences panel, which included Meselson, found the army's transportation plan to be woefully inadequate, and the disposal of weapons at sea (Operation CHASE, for "Cut Holes And Sink 'Em") had already resulted in an unintended explosion and other mishaps. As a result of the National Academy report, the army agreed to dispose of the weapons at the Rocky Mountain Arsenal, with the exception of the encased sarin rockets, which were dumped into the ocean off the coast of Florida.

In 1978, alarmed by local media reports and neighborhood chatter of spikes in birth defects and unusual illnesses, Lois Gibbs organized her Love Canal, New York, community to shut down her son's school and relocate neighborhood families.[525-27] The school had been constructed atop a site where Hooker Chemical Corporation, a subsidiary of Occidental Petroleum, had disposed of more than 20,000 tons of toxic waste before selling the property to the local school board for one dollar—and, in the process, transferring liability to an unqualified recipient. Gibbs was an unlikely spearhead for this movement, as she did not attend university and lacked relevant training. Nevertheless, she founded the Love Canal Homeowner's Association and led citizen efforts to hold industry and government accountable. In response to an engineer's presentation of the planned remediation, Gibbs said, "Excuse me, I'm just a dumb housewife. I'm not an expert. You're the expert. I'm just going to use a little common sense."[526] She then explained the faults in the engineering that was supposed to prevent contamination of the neighborhood. After two years of pressure, President Jimmy Carter stated on October 1, 1980, that all Love Canal families would be relocated and compensated for the loss of their homes. The fiasco motivated the passage of the Comprehensive Environmental Response, Compensation, and Liability Act of 1980, better known as the Superfund for the cleanup of contaminated sites.

The list of similar disasters is long, and often, as in the case of Love Canal, ordinary people galvanized public opposition to corporate neglect. An illustrative example is provided by Erin Brockovich, a legal clerk who discovered that residents near a compressor station operated by Pacific Gas & Electric Company in California were sickened by chromium-6 leaching from the facility.[425] This led to a record 1996 court settlement resulting from the largest class-action lawsuit related to pollution in US history.

Chemical tragedies that occurred in developing countries were often larger in scale and impact, due to lackadaisical safety standards and ineffective responses, both attributable to fewer financial resources and weak governance. For example, the Union Carbide pesticide plant in Bhopal, India, produced the insecticide Sevin (carbaryl, a carbamate originally marketed in 1956).[421] Due to an unnerving list of equipment and personnel errors, and following a string of smaller chemical accidents, the plant spewed a deadly gas cloud over Bhopal in 1984 that killed thousands and sickened tens of thousands of residents. Because of India's poverty, the risk of synthesizing highly toxic chemicals in a population center was tolerated, while standard safety precautions were simultaneously neglected.

Such an imposition of risk from pesticides onto residents of neighborhoods near a manufacturing facility is mirrored by the risks faced by farmworkers. When Carson alerted the world to the dangers of persistent organochlorine pesticides, governments around the world phased out their production and use. Other products, such as organophosphate pesticides, replaced the organochlorine pesticides. The organophosphates had the advantage of short residence time in the environment, but ironically, often the disadvantage of greater toxicity.[421] These chemicals posed a greater occupational risk than the more persistent chemicals, leading to illnesses and deaths among farmworkers and their families. An expression of this shift in risk to such a low-income group with little political recourse is found in the grape boycott movement of César Chávez in the 1980s.[528]

In the three decades following Carson's death, annual pesticide consumption in the United States doubled to more than 1 billion pounds.[421] By the 1990s, organophosphates made up more than half of this use. Carson's book succeeded in changing the political climate to one of caution regarding organochlorine pesticides, but despite her warnings about the dangers of organophosphates, their use continued to rise, along with severe damage to wildlife.

Other new categories of insecticides also entered the marketplace with great fanfare, including the pyrethroids, synthetic analogs to pyrethrum, and the neonicotinoids, systemic insecticides that are synthetic versions of nicotine.[421,529] Given that pyrethrum and nicotine were among the first widely used insecticides, the uses of which were limited by high cost, the development of synthetic versions held great promise. Pyrethroids were first synthesized in 1949 and then improved in the late 1960s; a commercially important version called permethrin was discovered in 1972.

Neonicotinoids were first marketed in the 1980s and proved to be immensely popular, especially when governments began restricting prevalent organophosphate insecticides.[421] In 2013, neonicotinoids supplanted organophosphates as the most used insecticides in the world. As with previous classes of insecticides, many insects quickly evolved resistance to neonicotinoids. And like previous classes of insecticides, neonicotinoids have affected the balance of nature, as they kill countless birds and are partially responsible for colony collapse disorder (the loss of honeybee and bumblebee colonies throughout the world and consequently the loss of the important pollination services bees provide). Thus, in a repeat performance and in an effort to protect crops from insect pests, chemists produced a class of insecticides that kills pollinators and predators of pests, and thereby puts at risk the very crop production that the insecticides were designed to protect.

Other two-word phrases with deep implications for life on Earth joined "silent spring" in the decades following the book's publication. The public soon worried about "acid rain," "nuclear winter," "ozone hole," and "global warming." The toxic chemical problem that Carson highlighted turned out to be an underestimate, rather than a hyperbolic rant. She was aware that this was the case; there was only so much that the public could digest in a single book. "The problem that I dealt with in *Silent Spring*," Carson said, "is not an isolated one. It is merely one part of a sorry whole—the reckless pollution of our living world with harmful and dangerous substances."[471]

Scientists discovered that many pesticides (including DDT) are toxic in part because they disrupt the endocrine system. However, added to the list of endocrine-disrupting compounds is a dizzying array of products, including many PCBs, cleaning solutions, makeup and personal care products, plasticizers, and flame retardants. Indeed, the list of toxic chemical compounds used in daily life seems endless. Humans and other living creatures are exposed to thousands of these toxic chemicals, which interact in unpredictable ways, leading to developmental abnormalities, delayed illness, and even health effects passed across generations. Carson anticipated much of this, stating, "There is something more than mere feminine intuition behind my concern about the possibility that our freewheeling use of pesticides may endanger generations yet unborn."[471]

The public identification of these problems, and the pursuit of solutions, began with pesticides and *Silent Spring*. Perhaps even more important, Rachel Carson inspired ordinary people to educate themselves as citizen-scientists and to overpower the neglect and corruption of governments and corporations. "Having recognized and defined our values," she wrote shortly before her death, "we must defend them without fear and without apology."[513]

What does the history of pesticides imply about the future of pesticides? New chemicals will emerge that are effective for a time, until pests evolve resistance. New techniques and technologies will dent pest populations, and thereby limit the scourge of deadly diseases and famine. The enticement of these new technologies will be irresistible and driven by necessity. Some, like tetraethyl lead and Freon, will result in serious unintended consequences. Others will vastly improve living conditions, and perhaps ease tensions and competition for resources and thereby reduce the risk of war. Some will be adapted for war, which seems to be an inevitable consequence of human existence. Corporations will profit, and tensions between corporations, government regulators, and ordinary people will play out in a political theater forever altered by the words of Rachel Carson:

It seems reasonable to believe that the more clearly we can focus our attention on the wonders and realities of the universe about us, the less taste we shall have for the destruction of our race. Wonder and humility are wholesome emotions, and they do not exist side by side with a lust for destruction.[447]

Epilogue

Ever since Friedrich Wöhler accidentally synthesized urea from cyanic acid and ammonia in 1828,[300] scientists have probed the workings of atoms and molecules to halt famine, to wage war on disease, and to destroy armies. Reflecting on this history, I find it rewarding to summarize my thoughts in the first person rather than the third, and to indulge in a rambling family history that explains my personal motivation to write about the interplay of scientists, chemistry, progress, and tragedy. Please forgive this change of style, as it allows me to relate a few stories of my ancestors and to think through the implications.

Earlier in the book, I wrote about the physicist James Franck—about his work with his mentor and friend Fritz Haber testing the efficacy of gas masks during World War I, about his resignation as institute director in protest of Nazi antisemitism, and about the strange story of how the future Nobel Laureate George de Hevesy dissolved Franck's Nobel gold medal along with that of Max von Laue using nitro-hydrochloric acid in Niels Bohr's institute in Copenhagen. But there is also a personal side to the story for me, because Franck was my great-grandfather. In our family, we call him Opa (Grandpa).

Before the Nazis rose to power, Opa visited Berkeley on a lecture circuit.[530] His daughter Dagmar, my grandmother, had a friend who was working in Berkeley on a Rockefeller fellowship from 1927 to 1928; that friend was Arthur von Hippel, who later became my grandfather. Arthur picked up Opa in San Francisco in the old Chevrolet that he had bought with a lab assistant for fifteen dollars.

In his autobiography, Arthur described what happened next:

> I met him at the train but, after leaving the railroad station, we found ourselves enveloped in the dense afternoon fog. I had not been in that area before and—turning at a street corner— found myself suddenly driving on a railroad embankment. Behind us an engine whistled and shoved off freight cars which came surging our way. Opa shouted that we should abandon the car; I shouted back that we could not afford to, that the car had cost $15, and jumped it down the embankment into a woodpile. There we landed with a crash but unhurt, and above us the ghostly silhouettes of the freight cars passed by. After bending the fenders straight, we reached Berkeley without further mishaps. This incident cemented our friendship. Opa enjoyed riding backwards up the mountains, invited me for dinner "accidentally" when he knew I had not eaten, and took me along on expeditions to the observatories on Mt. Hamilton and Mt. Wilson. Since it was the time of prohibition, we also went with Gilbert Lewis [the famous Berkeley chemist] to his favorite "speakeasy" . . . The Mt. Wilson excursion proved especially interesting due to our discussions with Professor Hubble, who was working at the time on the red-shifts of distant galaxies and his ideas of an expanding universe.[530]

Two years later, in 1930 and back in Germany, Arthur asked Opa for Dagmar's hand in marriage. Arthur wrote, "Opa warned me of the rising tide of anti-Semitism and the coming of the Nazis, but

I told him that I had taken my stand as an anti-Nazi and had written a counter-declaration for our Gilde and youth movement. He still believed I was very unwise but let Daggie decide."[530] Initially, Arthur's family reacted as Opa had predicted. "My father, my brothers and Olga [sister] were first stunned by my marriage to a Jew," wrote Arthur, "but then stood by me all the way. My more-distant relatives were scandalized and wanted a family council to intervene. Only the old general, Konrad von Hippel, commander of the army in the Balkans during World War I, wrote me a lovely note and took my side."[530]

Arthur had a couple of fruitful scientific years following his marriage to Dagmar, interacting with the likes of the Nobel Laureates Albert Einstein, Max Planck, Gustav Hertz, Fritz Haber, and Walther Nernst.[530] He and Opa socialized with Nernst, who showed off his newest invention, the electric piano, which Planck sat down to play. They also socialized with Haber, and on one occasion, Arthur and Dagmar visited Haber on his farm in South Germany. They borrowed Haber's Daimler and nearly crashed when its roof collapsed. This was an especially close call because Dagmar was pregnant with my father, Arndt.

Things soon descended into chaos for the family. Here I quote extensively from Arthur's autobiography to explain what happened next:

> The Rector of the University, a Professor of Agriculture and ardent Nazi, called a meeting of all the Faculty and declared the constitution of the University to be abolished. He asked us to look out the windows where Reichswehr [Realm Defense] and Nazi Storm Troopers were lined up to break any incipient resistance. The First Physics Institute under Professor Pohl joined the Nazis. Our Second Physics Institute [led by Franck] resisted, but found a traitor in its midst: one of our Ph.D. students turned out to be a Nazi leader who had hidden secret Nazi plans for the takeover in his cabinet.[530]

Believing that Arthur had accidentally found the secret plans, the student threatened Arthur with arrest. "Soon our personal lives became strongly affected by the fact that Daggie was Jewish. Old 'friends' suddenly appeared shortsighted and could not recognize us anymore. When I walked in the streets, people crossed over to the other side. Our father had to certify his 'Aryan origin.' Our East-Prussian uncle, Walther von Hippel, Chief-Officer of that Province and the family historian, who had been especially agitated by my marriage to Daggie, was thrown into jail by the Nazi Gauleiter [party governor] Koch, whom he had previously dismissed as incompetent. Our father defended Walther before the German Supreme Court and got him freed but the Nazis simply put him back into jail. Uncle Walther wrote me a letter of apology and then committed suicide."[530]

"In the spring of 1933, a Hitler edict banned Jewish students from the universities and Jewish professors, including [Max] Born [future Nobel Laureate] and [Richard] Courant [prominent mathematician], were subsequently dismissed. Opa Franck was exempted since he had received the Iron Cross, First Class, for heroism during World War I. Obviously, he did not want this preference. We therefore sat down with him and some friends to formulate a statement of resignation."[530] Opa's resignation letter to the government read: "With these lines I ask you, Mr. Minister, please to release me from my duties as regular professor at the University of Göttingen and director of Physics Institute II of this university. This decision is an intrinsic necessity for me because of the attitude of the government toward German Jewry. Most Sincerely, Prof. Dr. James Franck."[308]

Arthur wrote of their efforts to publicize Opa's resignation. "In the early morning hours, we telephoned the declaration to the *Göttinger Zeitung* [Göttingen News]."[530] The paper published the following account:

> The director of the Second Physical Institute of the University of Göttingen, Prof. James Franck, has requested from the Prussian minister of Sciences, Arts, and Cultural Affairs

immediate release from his official duties. This news will cause a major sensation not only in Göttingen, but throughout Germany, and one can even justifiably say, throughout the world. Franck is not just any lecturer of local or preeminent national importance. Franck's international reputation and global fame is unsurpassed by virtually any other German scholar today. When he received the Nobel prize a few years ago, the whole of Germany considered it an exceptional honor—because a German had again spread the fame of German scientific research beyond its borders. When such a man, who is only fifty, voluntarily relinquishes his teaching and research activities, the loss to science is beyond estimation.[308]

Lise Meitner, who later discovered fission with her colleague Otto Hahn, wrote to Opa, "Your dear letter naturally first gave me an inner jolt; but upon closer consideration and after I had read the wording of your letter to the rector, I have to concede that you are right. One cannot live contrary to one's convictions."[308] The prominent scientist Michael Polányi (whose son John won the 1986 Nobel Prize in Chemistry), wrote to Opa, "I learned about your step with astonishment and joy. As long as Jews exist, what you have done to save their honor will not be forgotten."[308] The Berlin rabbi Joachim Prinz wrote, "I feel it as my duty and a need to thank you for the extraordinary example you have offered in these difficult days for German Jewishness and German persons."[308] The Reich League of Jewish War Veterans wrote to Opa: "Esteemed fellow-soldier, Professor Dr. Franck, We have the strong desire to express to you our admiration and our thanks for your wonderful standpoint in your capacity as a front-line soldier and as a Jew. By it you have given German Jews moral support that has no equal. We are proud to be able to count you among our number."[308] News of Opa's resignation spread through the media in the United Kingdom, United States, the Netherlands, Italy, and elsewhere.

Arthur wrote, "Opa's wonderfully dignified statement of April 1933 came as a bombshell to the Nazis and the University faculty who had made peace with them. A counter-declaration condemning the statement appeared in the *Göttinger Tageblatt* [Göttingen Daily] of April 24."[530] This letter, signed by forty-two faculty members, stated, "We are in agreement that the form of the above tender of resignation is tantamount to an act of sabotage; and we therefore hope that the government will carry out the necessary purging measures expeditiously."[308] These signatories thereby gained the opportunity to replace Jewish faculty in more senior positions.

Arthur wrote, "Since the Nazis had already tapped our telephone lines, we were also individually attacked in the main Nazi newspaper, the *Völkischer Beobachter* [People's Observer]. I was so angry that I went to the Nazi-headquarters in Göttingen—characteristically located in the 'Jüdenstrasse' [Jew Street]—and tried to challenge its leader to a duel."[530] The Nazis then shut down the *Göttinger Zeitung* for publishing Opa's declaration.

Opa considered his professional options outside of civil service within Germany. He wrote to Max Born, "I told Planck that I would accept any position that was available that would offer me the opportunity to conduct research in Germany and some income, so long as it did not include any character of state employ. For I do not want to be a civil servant as long as the martial laws against Jews exist."[308] Opa inquired about potential employment with I. G. Farben via a position as a guest scientist at the Kaiser Wilhelm Institutes. Carl Bosch wanted to offer help, but the political realities prevented it. The Nobel Laureate and virulent antisemite Philipp Lenard, who was also a senator in the Nazi government, petitioned the Senate: "I submit the following three questions to the Senate in writing: Will the Senate support that (1) the Jew Fritz Haber, (2) the Jew James Franck, (3) the Jesuit Muckermann be removed immediately, respectively kept completely away from the institutes of the Kaiser Wilhelm Society?"[308]

Members of the National Socialist Student League raided bookstores and libraries, and forcibly removed all books by Jewish authors.[308] These they burned in giant bonfires in cities around the country beginning on May 10, 1933, including in Göttingen. The university's new rector spoke to the large crowd at the Göttingen book burning about the struggle against the "un-German spirit," a fight that was only just beginning, he said, with the bonfires. Germany no longer had a place for Opa.

Einstein wrote to Born at the end of May: "I am glad that you have resigned your positions (you and Franck). Thank God there is no risk involved for either of you. But my heart aches at the thought of the young ones. Lindemann [Frederick Alexander Lindemann, a physicist and advisor to Winston Churchill] has gone to Göttingen and Berlin (for one week). Maybe you could write to him there about [Edward] Teller. I heard that the establishment of a good Institute of Physics in Palestine (Jerusalem) is at present being considered. There has been a nasty mess there up to now, complete charlatanism. But if I get the impression that this business could be taken seriously, I shall write to you at once with further details."[308] Born relocated to Cambridge after an invitation from Nobel Laureate Ernest Rutherford.

Arthur wrote, "Professor Franck's declaration was re-published in England and Professor Lindeman[n] of Oxford came over to help. Lindeman[n] offered to take me back to Oxford but we felt that Heini Kuhn, the only one of us with a Jewish background, was more endangered. Heini and Mariele therefore went on to Oxford and a distinguished career. Soon thereafter, Professor Schwartz in Zurich succeeded in arranging with the Turkish dictator, the Ghazi Mustafa Kemal [Atatürk], that a new European-type university should be founded in Istanbul and about 30 European professors be hired to staff it. I was one of the 'lucky ones' selected . . . In our last night together at Göttingen, a tremendous display of shooting stars occurred. We watched in awe in the backyard with our friends, the Beyers, and took it as an omen of things to come."[530]

Arthur and Dagmar settled down in Istanbul with their boys Peter and Arndt. Arthur wrote about the events that followed: "I inherited a section of the old Sultan's palace as my future laboratory and the botanists, Heilbronn and Brauner, were installed in the former Mohammedan seminary . . . Two nights later we intentionally

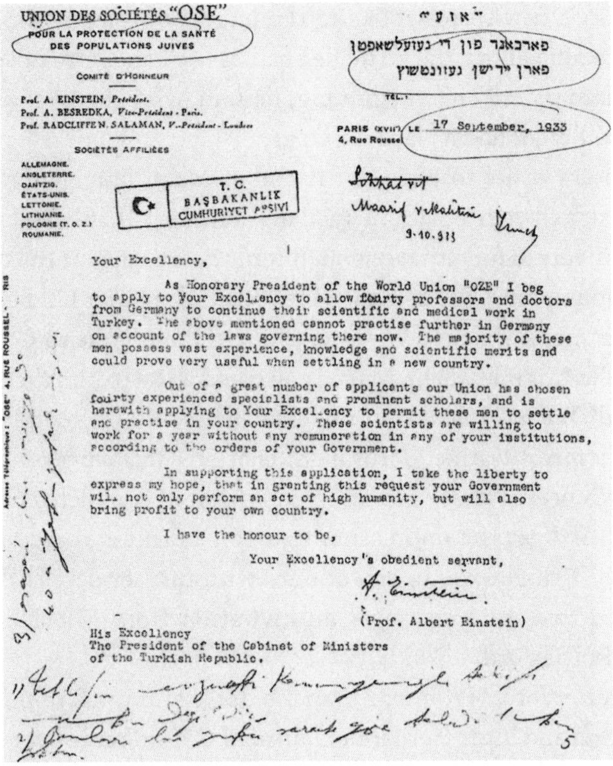

UNION DES SOCIÉTÉS "OSE"
POUR LA PROTECTION DE LA SANTÉ
DES POPULATIONS JUIVES

COMITÉ D'HONNEUR

Prof. A. EINSTEIN, Président.
Prof. A. BESREDKA, Vice-Président - Paris.
Prof. RADCLIFFE N. SALAMAN, V.-Président - Londres

SOCIÉTÉS AFFILIÉES

ALLEMAGNE.
ANGLETERRE.
DANTZIG.
ÉTATS-UNIS.
LETTONIE.
LITHUANIE.
POLOGNE (T. O. Z.)
ROUMANIE.

"אוזע"
פארבאנד פון די געזעלשאפטן
פארן יידישן געזונטשוץ

TEL.

PARIS (XVII°), LE 17 September, 1933
4, Rue Roussel

T. C.
BAŞBAKANLIK
CUMHURİYET ARŞİVİ

Your Excellency,

As Honorary President of the World Union "OZE" I beg to apply to Your Excellency to allow fourty professors and doctors from Germany to continue their scientific and medical work in Turkey. The above mentioned cannot practise further in Germany on account of the laws governing there now. The majority of these men possess vast experience, knowledge and scientific merits and could prove very useful when settling in a new country.

Out of a great number of applicants our Union has chosen fourty experienced specialists and prominent scholars, and is herewith applying to Your Excellency to permit these men to settle and practise in your country. These scientists are willing to work for a year without any remuneration in any of your institutions, according to the orders of your Government.

In supporting this application, I take the liberty to express my hope, that in granting this request your Government will not only perform an act of high humanity, but will also bring profit to your own country.

I have the honour to be,

Your Excellency's obedient servant,

A. Einstein

(Prof. Albert Einstein)

His Excellency
The President of the Cabinet of Ministers
of the Turkish Republic.

Fig. 13.1. Albert Einstein used his influence to help rescue Jewish professors by gaining for them employment in Turkey (and in the United States).[531-33] Among those saved was Arthur von Hippel, my grandfather. This is Einstein's letter, on behalf of a Paris-based Jewish relief organization, to Turkish Prime Minister Ismet Inonu, dated September 17, 1933. The notes on the letter, written by government officials, indicate that the scheme nearly failed. One note says, "This proposal is incompatible with clauses [in the existing laws]"; another says, "It is impossible to accept it due to prevailing conditions."[532] President Ataturk apparently made the decision to proceed. At the request of the Rockefeller Foundation, Opa and Richard Courant assessed the likelihood of success of Turkey's modernization of higher education. They reported that they found "a decided wish [among Turkish officials] to create a promising scientific center in Istanbul, which should contribute to the development of higher education in Turkey."[532] From collections of the Prime Ministry General Directorate of State Archives, Republic of Turkey.

missed a big festivity given for the foreign professors by the Gazi at the palace in Dolmabadge. We did so for a strange reason: the Gazi had the habit of absconding with any lady he liked especially well and of keeping her for a few days before returning her to her rightful husband."[530] Arthur and his mechanic then built a new physics lab using leftovers from old battleships and other odds and ends that they purchased in the bazaar.

Unfortunately, Turkish professors were fired to make room for these world-class scientists, and they met the arrival of the Germans with hostility and intrigue.[530] One Turkish former professor poisoned his German successor; fortunately, the victim survived. Other former faculty made accusations to Atatürk that the German professors were frauds, which led to formal inquiries.

Arthur, lecturing in German and French, found himself embroiled in turmoil when his interpreter translated his lecture on power generators as follows: "The professor says that he does not want to talk about the details of design because you are too foolish to understand it anyhow and had better buy these machines from abroad by planting potatoes and oranges in your country."[530] The students, outraged, shut down the university with a strike. Arthur wrote, "Honest misunderstanding or plot, the consequences seemed catastrophic: the Prime Minister and the Minister of Education came from Ankara; my colleagues trembled and mostly deserted me; the newspapers wrote articles that rather than being a professor before, I had sold old clothes; Daggie, taking the children for a walk, got commiserating news about her husband's dark past from charitable ladies of the neighborhood. Finally, our contract— originally drawn for five years—was shortened by mutual agreement to one year."[530]

Opa had already joined Bohr's institute in Copenhagen (where the physicist Edward Teller also relocated), working in a position funded by the Rockefeller Foundation, by the time Arthur's situation grew perilous in Turkey.[308] Opa and Bohr had a strong connection through their work. Opa had received the Nobel Prize in 1925 (along with Gustav Hertz) for the first experimental demonstration of Bohr's theory on atomic structure (for which Bohr had received the Nobel Prize in 1922). In 1930, when Teller and Opa first met, Opa told Teller, "Bohr's ideas did seem absurd, but Bohr was such a nice fellow that I felt he should at least be given a try."[534]

Opa began a long series of experiments investigating the physics of photosynthesis.[308] Feeling insecure about his findings in this new field, he corresponded with Richard Willstätter, who had deciphered the structure of chlorophyll in 1913. Additionally, along with de Hevesy and Bohr and others, Opa investigated radioactive decay.[530]

Opa felt preoccupied by Arthur and Dagmar's predicament. He wrote to his former student Heinrich Kuhn, whom Lindemann had brought to Oxford (in place of Arthur, at Arthur's request) and who would subsequently work on the Manhattan Project, "Whether or not we'll be staying here probably also depends on Hippel's fate. We've lost so much that we would at least like to have the children within reach."[308] Opa gave his Nobel Prize money to his two daughters in the hope that it would provide them with sufficient security in this perilous time.

At the invitation of Bohr, and to the great relief of Opa, Arthur relocated the family to Copenhagen to work in the institute from January 1935 until August 1936.[530] To celebrate the family reunion, Bohr lent his vacation cottage. Joining the family on the much-needed vacation was Teller, who subsequently also worked on the Manhattan Project and became known, to his distaste, as the "father of the hydrogen bomb."[534] Arthur wrote about this wonderful time working with Bohr: "In lectures, Bohr never quite knew in which language he spoke (Danish, German or English)—sometimes he switched in the midst of a thought. Then, he might suddenly stop altogether for awhile—his face going blank. Then came a beatific smile and a new idea had been born."[530]

At Bohr's request, Arthur made one last trip to Germany to acquire equipment for a high-voltage laboratory.[530] The lab was to enable nuclear excitation and disintegration experiments, which required a facility producing between 1 million and 2 million volts. Nernst had already conceived of such enormous electrical output with his thunderstorm generator on Monte Generoso in Switzerland, but that prototype was unsuitable for Bohr's requirements. Arthur was to visit a German company producing a new cascaded transformer capable of 2 million volts.

On that sorrowful visit, Arthur learned that many of his old friends were now Nazis, and he saw Hitler himself driving by.[530] Someone in the German embassy in Copenhagen warned Dagmar that the Nazis intended to arrest Arthur, and Dagmar relayed this message to Arthur via a friend. Arthur escaped from Germany by taking a flight back to Denmark, rather than the planned train where Nazi enforcers waited for his arrival.

The positions in Bohr's institute were temporary, and the family prepared for a new life in America. As he readied to emigrate, Opa invited Planck to join him for a few days in Denmark. Planck replied, "No, I cannot travel abroad. On my previous travels I felt myself to be a representative of German science and was proud of it—now I would have to hide my face in shame."[308]

Opa was the first in the family to arrive in America, where he accepted a position at Johns Hopkins University and set up the practicalities of a new life before the others came.[308] He then returned to Europe in the summer of 1936 to help his daughters, their husbands, and his grandchildren to emigrate.

At the end of August 1936, Bohr and his family delivered Opa, Arthur, and Dagmar, along with the two boys, to their ship, the *Scanstates*, which then set sail for America. The physicist Karl Compton, who was the president of MIT (and had previously been a visiting professor at Opa's institute in Göttingen, and coincidentally was a member of a previous generation of my wife's family), had offered Arthur a position there.[313] The family settled into their American apartment, and immediately the upstairs neighbors invited my father and Uncle Peter to a birthday party for their little girl. The girl said to Peter, "My ancestors came over on the *Mayflower*!" Peter responded, "My ancestors came over on the *Scanstates*!"[530]

Both Arthur and Opa had fought for the German Army in the trenches in World War I.[530] When World War II broke out, they both

Fig. 13.2. Opa Franck, Uncle Peter (on Opa's right), and my father,
Arndt (on Opa's left) depart Copenhagen on the *Scanstates*.
This photo was published in the newspaper *Ekstrabladet*. US officials
processed the family's immigration papers on Ellis Island.

turned their scientific aptitudes into potent tools for the US military in the fight against Nazi Germany. With the blessing of Karl Compton and a $5,000 grant, Arthur created the MIT Laboratory for Insulation Research. It became one of the largest materials research laboratories in support of the military effort. Arthur and his group developed the dielectric materials for radar, and he worked simultaneously in the Radiation Laboratory at MIT to integrate these materials into the new radar technology. His lab also produced various plastics, rubbers, ceramics, crystals, and other materials for a variety of wartime technologies and improved materials to enhance performance of semiconductors and photocells.

Arthur's Laboratory for Insulation Research formed a collaboration called the War Committee on Dielectrics with the army, the navy, and the War Production Board.[530] The government tasked this committee to solve technical problems with materials experienced in the war theaters. One of these arose in New Guinea, where mites and fungi plagued Allied forces by devouring their uniforms and equipment. Arthur proposed an effective solution of replacing compromised materials with halogenated compounds. Due to existing contracts with suppliers, the military delayed this solution, but eventually it used the polyvinyl chloride suggested by Arthur, which solved the pest problem. Arthur wrote, "The later abuse of such compounds in the form of sprays here at home, with the resulting threat of Miss Carson's 'Silent Spring,' could not then be foreseen."[530] For his wartime work, Arthur received the US President's Certificate of Merit from President Truman in 1948.[535]

Meanwhile, Opa moved from Johns Hopkins to the University of Chicago, where he collaborated with Teller on the study of photosynthesis.[308] But with the outbreak of war, they turned their efforts from the basic science they loved to the practical problem of developing nuclear weapons before the Nazis did. Robert Oppenheimer directed the effort. Many years earlier, Opa had attended Oppenheimer's PhD exam in Göttingen; Oppenheimer said afterward, "I got out of there just in time. He was beginning to ask questions."[314] The Nobel Laureate Enrico Fermi constructed a uranium reactor in an old squash court at the university, and the Nobel Laureate Arthur Holly Compton (Karl Compton's brother and also a member of a previous generation of my wife's family) directed the research in Chicago.[308] In another coincidence of my family's history, Arthur Holly Compton appointed Opa to lead the chemistry division of the nuclear bomb research. Opa agreed, on the condition that once the bomb was ready, he would be permitted to provide his opinion about its use to a senior policymaker.[313] Opa feared that the Nazis would obtain the bomb first, thereby ensuring their victory.[314] He also feared the consequences of US governmental control of science, having lived through the Nazi consolidation of such control in Germany.

The bad news from Europe arrived relentlessly. Lise Meitner escaped to Sweden, and wrote to Opa of the fate of their friends back in Germany.[308] The Germans executed Planck's son Erwin for his role in the failed assassination attempt on Hitler, and similar grim news arrived with a depressing constancy. Opa wrote to Meitner that he yearned to be cheered up by Bohr; "I hope to see him during the summer and hope to become a little infected by his optimism and constructive attitude towards life."[308] Bohr, like Meitner, had escaped to Sweden.

The war in Europe ended with the defeat of Nazi Germany, and the scientists of the Manhattan Project turned their attention to the repercussions of this new weapon were it to be used against Japan.[308] Opa expressed his concerns and those of the other scientists to the secretary of commerce, Henry Wallace, "They cannot help but worry about the fact that mankind has learned to unleash atomic power without being ethically and politically prepared to use it wisely."[314] On June 5, 1945, Opa wrote a memorandum to the leadership of the Manhattan Project about political implications of the bomb. It read, in part, "We believe a bomb able to produce a sensational destruction will be available very soon. It took the United States 3 ½ years to reach that goal, and great sacrifices in the wealth of the nation had to be made for this progress and great scientific and industrial organizations were needed."[313]

The next day, Compton assigned to Opa the leadership of the committee to write about the social and political implications.[313] Other committee members included Glenn Seaborg (who subsequently chaired the Atomic Energy Commission), Leo Szilard (who had conceived the idea of nuclear chain reactions in 1933, but felt distressed over the repercussions), and Eugene Rabinowitch (who subsequently cofounded the *Bulletin of the Atomic Scientists*). The committee's report, completed in five days, became known as the "Franck Report." Opa, Compton, and the physicist Norman Hilberry attempted to deliver the report to Secretary of War Henry L. Stimson in Washington, DC. Stimson was absent from the city, and they left the report with his assistant, along with a note written by

Compton stating that, in his view, the report did not adequately consider the lives that would be saved if using the bomb on Japan hastened the end of the war.[313,536] Compton's note was based upon the analysis of Enrico Fermi, Ernest Orlando Lawrence, and Robert Oppenheimer, who concluded on June 16, "we cannot propose a technical demonstration likely to bring an end to the war, and we see no acceptable alternative to direct military use."[314]

The summary of the "Franck Report" reads as follows:

The development of nuclear power not only constitutes an important addition to the technological and military power of the United States, but also creates grave political and economic problems for the future of this country. Nuclear bombs cannot possibly remain a "secret weapon" at the exclusive disposal of this country for more than a few years. The scientific facts on which their construction is based are well known to scientists of other countries. Unless an effective international control of nuclear explosives is instituted, a race for nuclear armaments is certain to ensue following the first revelation of our possession of nuclear weapons to the world. Within ten years other countries may have nuclear bombs, each of which, weighing less than a ton, could destroy an urban area of more than ten square miles . . . We believe that these considerations make the use of nuclear bombs for an early unannounced attack against Japan inadvisable. If the United States were to be the first to release this new means of indiscriminate destruction upon mankind, she would sacrifice public support throughout the world, precipitate the race for armaments, and prejudice the possibility of reaching an international agreement on the future control of such weapons. Much more favorable conditions for the eventual achievement of such an agreement could be created if nuclear bombs were first revealed to the world by a demonstration in an appropriately selected uninhabited area . . . To sum up, we urge that the use of nuclear bombs in this war be considered as a problem of long-range national policy rather than of military expediency, and that this policy be directed primarily to the achievement of an agreement permitting an effective international control of the means of nuclear warfare.[537]

The "Franck Report," of course, did not change policy. Stimson rejected its recommendations on June 21, and President Truman never saw it.[314] On August 6, the United States dropped the uranium bomb over Hiroshima, and three days later dropped the plutonium bomb over Nagasaki. With this dramatic demonstration of the power of physics and chemistry, the war came to an end.

Opa turned his attention to humanitarian relief in Germany in a political climate clamoring for revenge.[308] He and other German exiles drafted a lengthy appeal to the American public to prevent a looming famine in Germany. The text read, in part, "We the undersigned—who were victims of this principle [the Nazi ideology] in Germany or who staked our existence on the fight against it—appeal to the American people to stand by the principle of justice. And we appeal to them in the name of charity. Many of us have barely escaped death; all of us have lost relatives or friends before the firing squads or in the torture camps of Hitler. For the last twelve years we have been haunted by the vision of helpless innocence and unavenged brutality. That vision is before us again today."[308]

Opa tried to convince Einstein to sign the appeal, but Einstein refused. Their letters debating the matter went back and forth. Einstein wrote in his final refusal:

Dear Franck,

I still remember the Germans' "campaign of tears" after the last war far too well to fall for that again. The Germans slaughtered millions of civilians according to a carefully conceived plan in order to steal their places. They would do it again if they could. The few white ravens among them changes absolutely nothing. From the few letters I have received from there I see that among the Germans there isn't a trace of remorse. I also see very clearly that the catering to the Germans has started all over again at the "United Nations"; these trends, the motivation behind the nursing of Germany back to strength after 1918, are most vibrant among the English; for concern about one's precious purse is stronger there, too, than any worries about one's dear fatherland. Dear Franck! Keep your hands off this foul affair! After abusing your kind-heartedness, they will make fun of your gullibility. But even if you can't be saved, I in any case will have nothing to do with the matter. And if a suitable opportunity presents itself, I will speak out against it in public.

Warm greetings to you, yours,
A. Einsten.[308]

This response did not break their friendship, and Opa stopped participating in the public appeal for aid to Germany.[308] He wrote to Max Born that he would prefer to avoid politics entirely, "if only my conscience would not force me to take a stand on a few political issues. I hate to be involved in anything political; I hate publicity, but I just cannot retire into the ivory tower of free research and forget about the world. And, of course, at our age we are probably more pessimistic than the young people. Even I am not consistent in my pessimistic point of view, because I have an elementary joy in each new grandchild, and feel that whenever I have the opportunity I am a kind of professional grandfather."[313]

Opa, along with Dagmar and Arthur, continued vigorous private efforts to send food and money to friends and relatives, and to help free those who were trapped in Russia or imprisoned elsewhere.[314,530,535] In a postwar speech in Hamburg and in the atmosphere of Stalin's power politics, Opa said, "We know that a large proportion of humanity is ruled by a dictatorship that opposes everything we regard as worthy of human beings, that enforces blind obedience by brutal means and is intent on establishing the rigid system of an ant state on the whole of mankind. In reward it promises heaven on earth as soon as all resistance against its pseudoreligion has been broken."[308]

In 1947, the German government offered Opa the chair in experimental physics in Heidelberg.[313] Opa declined. America was now the home of his family. He responded to the offer, "I believe I know that the majority of Germans rejected the murders committed against the Jews and the other races that the Nazis labeled inferior. And I do not reproach those people for not throwing themselves down the Moloch's gullet because they deemed it useless. But another considerable percentage of the populace stood by and watched the crimes with indifference. With them I want no contact. So I cannot imagine a fruitful teaching position in which I would have to ask myself whether this or that one with whom I had official or personal business was one of those."[313]

Despite his refusal to accept the Heidelberg chair, Opa reconciled with postwar Germany.[313] The Kaiser Wilhelm Society, renamed the Max Planck Society, offered him corresponding membership in 1948. Opa's friend Otto Hahn was now its president, and Opa accepted. In 1951, Opa and Hertz received the Max Planck Medal of the German Physical Society. Two years later, Opa received an honorary doctorate from the University of Heidelberg.[314] Also, along with Born and Courant, Opa accepted honorary citizenship from Göttingen—something they believed would honor the victims of the Nazis.[313] On May 21, 1964, while visiting Hahn and Born in Göttingen, Opa passed away.

These few anecdotes allow me to relate how the story of twentieth-century chemistry holds purchase in my mind. My family was deeply involved in all of it, from the gas warfare of World War I to the nuclear warfare of World War II. I grew up hearing stories about Haber and Einstein and most of the other twentieth-century scientists profiled in this book, about trench warfare with gas weapons, about the innocent beginnings and evil ends of I. G. Farben, about the rise of Nazism and immigration to America. The stories of Opa and Arthur, my great-grandfather and grandfather, respectively, allow me to comprehend the world events that tossed my family and millions of others about like a dinghy in a storm. That storm included genocide, famine, and world wars. It included the ravages of vector-borne diseases, and the life of a refugee.

Such events were not limited to my father's family. My mother was born in Vienna. Her parents were prominent psychoanalysts in Freud's group, and as Jews they, too, were desperate to flee following the Anschluss in 1938. With the help of a colleague in America, they escaped the fate of most European Jews.

I was born in Alaska three years after Rachel Carson died. We lived in a modest home adjacent to a forest that was, at the time, the edge of town. We had no television, and my three siblings and I roamed feral in the woods with all the other neighborhood kids. The air was clean, the water pure, and we raised our own animals on our family's small farm. But these early years of my life were also my first introduction to the passions raised by indiscriminate use of pesticides.

One summer day, as we played in the yard, a cloud of gas drifted in the breeze from our neighbor's land. In an effort to kill aphids, our neighbor had hired an exterminator. My father strapped his .44 revolver to his belt and threatened to use it should the exterminator continue spraying. The exterminator fled. This behavior continued with other exterminators until none in town were willing to spray our neighbor's yard. It was an elegant solution, but likely limited in its viability to that predigital age and to a frontier, such as Alaska, that still tolerated threats of gun violence.

My father thereby stimulated in me an early interest in pesticides. If a small cloud of gas could initiate such a ruckus, then surely the gas signified something important. I suspect that our neighbor's use of poisonous gas to kill aphids hit an especially raw nerve for my father, who, after all, had lived in four countries by the time he was five simply because his mother was Jewish. Relatives on both sides of our family had been killed in Nazi gas chambers. In that light, his reaction to protect his children from such a poison, especially coming on the heels of *Silent Spring*, seems perfectly reasonable. Indeed, people throughout the world reacted emotionally to *Silent Spring*, as well they should. After all, what could be more emotional than the health of one's children and the place of humanity in the balance of nature?

Acknowledgments

Optimism reigned in September 2011 when I signed my contract with the University of Chicago Press to write this book in eighteen months. Eighteen months stretched into eight years. I would like to thank my editor during this time, Christie Henry, for her patience. I also thank Christie for her initial interest in this project and her help in establishing a tone with the first chapter. My brother Bill von Hippel, my wife Cathy von Hippel, and my uncle Frank N. von Hippel all provided helpful comments on the book in draft form. I also thank two anonymous reviewers for their thoughtful comments. Mary DeJong, science librarian at Northern Arizona University, helped me track down difficult source material, such as recently declassified documents. I thank my new editor, Scott Gast, for stylistic improvements and for his efforts in shepherding this book to print. I also thank Lys Weiss, PhD, of Post Hoc Academic Publishing Services for her careful copyediting of the manuscript.

Map of Place Names

Map of place names. Created by Max von Hippel.

Barents Sea

Norway

B

Russia (Soviet Union)

Moscow

Ukraine Stalingrad

Caffa, Sebastopol, Crimea

Constantinople (Istanbul), Turkey   Assyria

Mediterranean Sea   Baghdad   Korea   Japan

Tunis, Tunisia   Aleppo   China   Hiroshima

Morocco   Jerusalem, Palestine (Israel)   Afghanistan   Nagasaki

Algeria   Cairo   Yunan Province   Canton

Egypt   Qatar   Karachi   Hanoi   Hong Kong

Arabia   Bhopal Calcutta   Laos   French Indochina   Wake Island

Bombay   Burma   Manila   Saipan Island

Bangalore   India   Cambodia   Bataan Guam   Bikini Atoll

Sierra Leone   Nigeria   Ethiopia   Nilgherry Hills   Suoi Dau, Nha Trang   Vietnam Mindanao   Philippines

Accra, the Gold Coast (Ghana)   Ceylon (Sri Lanka)   Malaya

Kenya   Singapore

German East Africa   Dutch East Indies   New Guinea

Java   Guadalcanal

Christmas Island

Madagascar

Australia

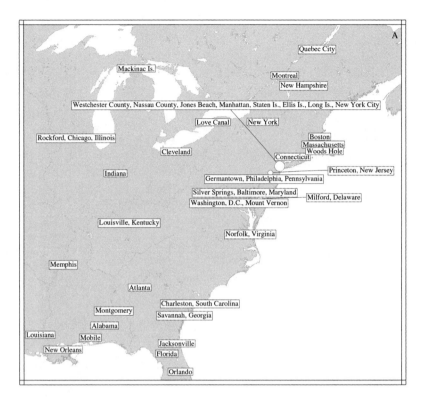

A

Quebec City

Mackinac Is.

Montreal

New Hampshire

Westchester County, Nassau County, Jones Beach, Manhattan, Staten Is., Ellis Is., Long Is., New York City

Love Canal   New York

Rockford, Chicago, Illinois

Cleveland

Boston
Massachusetts
Woods Hole
Connecticut

Princeton, New Jersey

Indiana

Germantown, Philadelphia, Pennsylvania

Silver Springs, Baltimore, Maryland

Milford, Delaware

Washington, D.C., Mount Vernon

Louisville, Kentucky

Norfolk, Virginia

Memphis

Atlanta

Charleston, South Carolina

Montgomery

Savannah, Georgia

Alabama

Louisiana   Mobile

New Orleans

Jacksonville

Florida

Orlando

Literature Cited

1   Loeb, A. P. Birth of the Kettering Doctrine: Fordism, Sloanism and the discovery of tetraethyl lead. *Business and Economic History* **24**, 72–87 (1995).

2   Thomas Midgley, Jr., American chemical engineer. *Encyclopædia Britannica.* https://www.britannica.com/biography/Thomas-Midgley-Jr (2018).

3   Needleman, H. L. The removal of lead from gasoline: historical and personal reflections. *Environmental Research Section A* **84**, 20–35 (2000).

4   McNeill, J. R. *Something New under the Sun: An Environmental History of the Twentieth-Century World.* (W. W. Norton & Co., 2000).

5   Hernberg, S. Lead poisoning in a historical perspective. *American Journal of Industrial Medicine* **38**, 244–54 (2000).

6   Byers, R. K., & Lord, E. E. Late effects of lead poisoning on mental development. *American Journal of Diseases of Children* **66**, 471–94 (1943).

7   Nevin, R. How lead exposure relates to temporal changes in IQ, violent crime, and unwed pregnancy. *Environmental Research* **83**, 1–22 (2000).

8   Nevin, R. Understanding international crime trends: the legacy of preschool lead exposure. *Environmental Research* **104**, 315–36 (2007).

9   Needleman, H. L., McFarland, C., Ness, R. B., Fienberg, S. E., & Tobin, M. J. Bone lead levels in adjudicated delinquents: a case control study. *Neurotoxicology and Teratology* **24**, 711–17 (2002).

10 Wright, J. P., et al. Association of prenatal and childhood blood lead concentrations with criminal arrests in early adulthood. *PLoS Medicine* **5**, e101 (2008).

11 Fergusson, D. M., Boden, J. M., & Horwood, L. J. Dentine lead levels in childhood and criminal behaviour in late adolescence and early adulthood. *Journal of Epidemiology & Community Health* **62**, 1045–50 (2008).

12   Hall, W. Did the elimination of lead from petrol reduce crime in the USA in the 1990s? *F1000Research* **2**, 156 (2013).

13   Reyes, J. W. Environmental policy as social policy? The impact of childhood lead exposure on crime. *B. E. Journal of Economic Analysis & Policy* **7** (2007).

14   Boutwell, B. B., et al. The intersection of aggregate-level lead exposure and crime. *Environmental Research* **148**, 79–85 (2016).

15   Mielke, H. W., & Zahran, S. The urban rise and fall of air lead (Pb) and the latent surge and retreat of societal violence. *Environment International* **43**, 48–55 (2012).

16   Thompson, R. J. Freon, a refrigerant. *Industrial and Engineering Chemistry* **24**, 620–23 (1932).

17   Wang, L. 1941: Thomas Midgley Jr. (1889–1944). *Chemical and Engineering News* **86** (2008).

18   Molina, M. J., & Rowland, F. S. Stratospheric sink for chlorofluoromethanes: chlorine atom-catalysed destruction of ozone. *Nature* **249**, 810–12 (1974).

19   The Nobel Prize in Chemistry 1995. NobelPrize.org. https://www.nobelprize .org/prizes/chemistry/1995/summary/ (1995).

20   Ramanathan, V. Greenhouse effect due to chlorofluorocarbons: climatic implications. *Science* **190**, 50–52 (1975).

21   O'Rourke, J. *The History of the Great Irish Famine of 1847, with Notices of Earlier Irish Famines.* (James Duffy & Co., Ltd., 1902).

22   *Lost Crops of the Incas.* (National Research Council, National Academy Press, 1989).

23   Grubb, E. H., & Guilford, W. S. *The Potato: A Compilation of Information from Every Available Source.* (Doubleday, Page & Co., 1912).

24   Wright, W. P., & Castle, E. J. *Pictorial Practical Potato Growing.* (Cassell & Co., Ltd., 1906).

25   Warolin, C. Homage to Antoine-Augustin Parmentier (1737–1813), first president of the Pharmacy Society of Paris in 1803. *Annales pharmaceutiques françaises* **63**, 340–42 (2005).

26   Block, B. P. Antoine-Augustin Parmentier: pharmacist extraordinaire. *Pharmaceutical Historian* **38**, 6–14 (2008).

27   Woodham-Smith, C. *The Great Hunger, Ireland 1845–1849.* (Hamish Hamilton, 1962).

28   Andrivon, D. The origin of *Phytophthora infestans* populations present in Europe in the 1840s: a critical review of historical and scientific evidence. *Plant Pathology* **45**, 1027–35 (1996).

29   Gibbs, C. R. V. *Passenger Liners of the Western Ocean: A Record of North Atlantic Steam and Motor Passenger Vessels from 1838 to the Present Day.* (Staples Press, 1952).

30   Jones, L. R., Giddings, N. J. & Lutman, B. F. Investigations on the potato fungus *Phytophthora infestans*. *Vermont Agricultural Experiment Station Bulletin* **168** (1912).

31   Jensen, J. L. Moyens de combattre et de détruire le *Peronospora* de la pomme de terre. *Mémoires Société Nationale d'Agriculture de France* **131**, 31–156 (1887).

32   Trevelyan, C. E. *The Irish Crisis*. (Longman, Brown, Green & Longmans, 1848).

33   Berkeley, M. J. Observations, botanical and physiological, on the potato murrain. *Journal of the Horticultural Society of London* **1**, 9–34 (1846).

34   Solly, E. Chemical observations on the cause of potato murrain. *Journal of the Horticultural Society of London* **1**, 35–42 (1846).

35   Townley, J. *The Potato*. (Benjamin Lepard Green, 1847).

36   Large, E. C. *The Advance of the Fungi*. (Jonathan Cape, 1940).

37   Fabricius, J. C. Forsøg til en abhandling om planternes sygdomme. *Det Kongelige Norske Videnskabers Selskabs Skrifter* **5**, 431–92 (1774).

38   Whetzel, H. H. *An Outline of the History of Phytopathology*. (W. B. Saunders Co., 1918).

39   Vallery-Radot, R. *Louis Pasteur, His Life and Labours*. (D. Appleton & Co., 1885).

40   Zinsser, H. *Rats, Lice and History: Being a Study in Biography, which, after Twelve Preliminary Chapters Indispensable for the Preparation of the Lay Reader, Deals with the Life History of Typhus Fever*. (Little, Brown & Co., 1934).

41   Darwin, C. *On the Origin of Species by Means of Natural Selection, or the Preservation of Favoured Races in the Struggle for Life*. (John Murray, 1859).

42   Dubos, R. J. *Louis Pasteur, Free Lance of Science*. (Little, Brown & Co., 1950).

43   Ullmann, A. Pasteur-Koch: distinctive ways of thinking about infectious diseases. *Microbe* **2**, 383–87 (2007).

44   Lister, J. On the antiseptic principle of the practice of surgery. *British Medical Journal* **21, September**, 246–48 (1867).

45   De Bary, H. A. *Die gegenwärtig herrschende Kartoffelkrankheit, ihre Ursache und ihre Verhütung*. (Förstner, 1861).

46   Margulis, L., Corliss, J. O., Melkonian, M., & Chapman, D. J. *Handbook of Protoctista*. (Jones & Bartlett, 1990).

47   Compton, D. A. *Potato Culture*. (Orange Judd Co., 1870).

48   Millardet, P. M. A. Traitement du mildiou et du rot. *Journal d'agriculture pratique* **2**, 513–16 (1885). Trans. Felix John Schneiderhan, Phytopathological Classics 3, 7–11. Ithaca, NY: Cayuga Press for American Phytopathological Society, 1933.

49   Gayon, U., & Sauvageau, C. Notice sur la vie et les travaux de A. Millardet. *Mémoires de la Société des Sciences Physiques et Naturelles de Bordeaux* **6**, 9–47 (1903).

50 Schneiderhan, F. J. Pierre Marie Alexis Millardet. *Phytopathological Classics* **3**, 4 (1933).

51 Ayres, P. G. Alexis Millardet: France's forgotten mycologist. *Mycologist* **18**, 23–26 (2004).

52 Millardet, P. M. A. Traitement du mildiou par le mélange de sulphate de cuivre et de chaux. *Journal d'agriculture pratique* **2**, 707–10 (1885). Trans. Felix John Schneiderhan, Phytopathological Classics 3, 12–17. Ithaca, NY: Cayuga Press for American Phytopathological Society, 1933.

53 Millardet, P. M. A. Sur l'histoire du traitement du mildiou par le sulfate de cuivre. *Journal d'agriculture pratique* **2**, 801–5 (1885). Trans. Felix John Schneiderhan, Phytopathological Classics 3, 18–25. Ithaca, NY: Cayuga Press for American Phytopathological Society, 1933.

54 King, A. F. A. Insects and disease: mosquitoes and malaria. *Popular Science Monthly* **23** (1883).

55 Cox, F. E. G. History of human parasitology. *Clinical Microbiology Reviews* **15**, 595–612 (2002).

56 Webb, J. L. A., Jr. *Humanity's Burden: A Global History of Malaria.* (Cambridge University Press, 2009).

57 Hippocrates. *Of the Epidemics* (trans. Francis Adams). (400 BCE). In *The Genuine Works of Hippocrates, Translated from the Greek with a Preliminary Discourse and Annotations by Francis Adams.* (Printed for the Sydenham Society, C. & J. Adlard, Printers, 1849).

58 Manson-Bahr, P. The jubilee of Sir Patrick Manson (1878–1938): a tribute to his work on the malaria problem. *Post-Graduate Medical Journal* **November**, 345–57 (1938).

59 Shakespeare, W. *The Tempest.* (Isaac Iaggard & Ed. Blount, 1623). In *The Works of William Shakespeare, the Text Revised by the Rev. Alexander Dyce, In Ten Volumes*, vol. 1. 4th ed. (Bickers & Son, 1880).

60 Hempelmann, E., & Krafts, K. Bad air, amulets and mosquitoes: 2,000 years of changing perspectives on malaria. *Malaria Journal* **12**, 1–13 (2013).

61 Duffy, J. *Epidemics in Colonial America.* (Kennicat Press, 1972).

62 Melville, C. H. The prevention of malaria in war. In *The Prevention of Malaria* (ed. R. Ross). (John Murray, 1910).

63 Russell, P. F. Introduction. In *Preventive Medicine in World War II, Vol. 6: Communicable Diseases: Malaria* (ed. J. Boyd). (Office of the Surgeon General, Department of the Army, 1963).

64 Malaria and the progress of medicine. *Popular Science Monthly* **24**, 238–43 (1884).

65 Ross, R. Researches on malaria. Nobel lecture, **December 12**. (1902).

66 Torti, F. *Therapeutice specialis ad febres quasdam periodicas perniciosas.* (B. Soliani, 1712).

67   Jackson, R. *A Treatise on the Fevers of Jamaica, with Some Observations on the Intermitting Fever of America, and an Appendix, Containing Some Hints on the Means of Preserving the Health of Soldiers in Hot Climates.* (J. Murray, 1791).

68   Pelletier, P. J., & Caventou, J. B. *Recherches chimiques sur les quinquinas.* (Crochard, 1820).

69   *Nobel Lectures, Physiology or Medicine, 1901–1921.* (Elsevier Publishing Co., 1967).

70   Laveran, A. Note sur un nouveau parasite trouvé dans le sang de plusieurs malades atteints de fièvre palustre. *Bulletin de l Académie Nationale de Medicine (Paris)* **9**, 1235–36 (1880).

71   Laveran, A. *Traité des fièvres palustres avec la description des microbes du paludisme.* (Octave Doin, 1884).

72   Manson, P. On the development of *Filaria sanguinis hominis* and on the mosquito considered as a nurse. *Journal of the Linnean Society (Zoology)* **14**, 304–11 (1878).

73   Manson-Bahr, P. H., & Alcock, A. *The Life and Work of Sir Patrick Manson.* (Cassell & Co., Ltd., 1927).

74   Manson, P. On the nature and significance of crescentic and flagellated bodies in malarial blood. *British Medical Journal* **2**, 1306–8 (1894).

75   Nuttall, G. H. F. On the role of insects, arachnids and myriapods, as carriers in the spread of bacterial and parasitic diseases of man and animals: a critical and historical study. *Johns Hopkins Hospital Reports* **8**, 1–155 (1899).

76   Roos, C. A. Physicians to the presidents, and their patients: a biobibliography. *Bulletin of the Medical Library Association* **49**, 291–360 (1961).

77   Howard, L. O. Dr. A. F. A. King on mosquitoes and malaria. *Science* **41**, 312–15 (1915).

78   Ross, R. *The Prevention of Malaria.* (E. P. Dutton & Co., 1910).

79   Bynum, W. The art of medicine: experimenting with fire: giving malaria. *Lancet* **376**, 1534–35 (2010).

80   Ross, R. Fever with intestinal lesions. *Transactions of the South Indian Branch of the British Medical Association* (1892).

81   Ross, R. Cases of febricula with abdominal tenderness. *Indian Medical Gazette*, 166 (1892).

82   Ross, R. Entero-septic fevers. *Indian Medical Gazette*, 230 (1892).

83   Ross, R. A study of Indian fevers. *Indian Medical Gazette*, 290 (1892).

84   Ross, R. Some observations on haematozoic theories of malaria. *Medical Reporter*, 65 (1893).

85   Ross, R. Inaugural lecture on the possibility of extirpating malaria from certain localities by a new method. *British Medical Journal* **July 1**, 1–4 (1899).

86   Ross, R. Observations on the crescent-sphere flagella metamorphosis of the malarial parasite within the mosquito. *Transactions of the South Indian Branch of the British Medical Association* **December** (1895).

87   Ross, R. Some experiments in the production of malarial fever by means of the mosquito. *Transactions of the South Indian Branch of the British Medical Association* (1896).

88   Guillemin, J. Choosing scientific patrimony: Sir Ronald Ross, Alphonse Laveran, and the mosquito-vector hypothesis for malaria. *Journal of the History of Medicine and Allied Sciences* **57**, 385–409 (2002).

89   Ross, R. On some peculiar pigmented cells found in two mosquitoes fed on malarial blood. *British Medical Journal*, 1786 (1897).

90   Manson, P. Surgeon-Major Ronald Ross's recent investigations on the mosquito-malaria theory. *British Medical Journal* **June 18**, 1575–77 (1898).

91   Ross, R. *Preliminary Report on the Infection of Birds with Proteosoma by the Bites of Mosquitoes.* (Government Press, 1898).

92   Manson, P. The mosquito and the malaria parasite. *British Medical Journal* **2**, 849–53 (1898).

93   Bignami. Come si prendone le febri malariche. *Bull. Accad. Med. Roma* **November 15** (1898). Translation: The inoculation theory of malarial infection: account of a successful experiment with mosquitoes. *Lancet* **152**, 1461–63 (1898).

94   Manson, P. Experimental proof of the mosquito-malaria theory. *British Medical Journal* **2**, 949–51 (1900).

95   Manson, P. T. Experimental malaria: recurrence after nine months. *British Medical Journal* **July 13**, 77 (1901).

96   G. H. F. N. In memoriam: Patrick Thurburn Manson. *Journal of Hygiene* **2**, 382–83 (1902).

97   Marotel, G. The relation of mosquitoes, flies, ticks, fleas, and other arthropods to pathology. *United States Congressional Serial Set, Annual Report Smithsonian Institution, 1909*, 703–22 (1910).

98   Koch, R. Zweiter Bericht über die Thatigkeit der Malaria-Expedition. *Deutsche medizinische Wochenschrift* **26**, 88–90 (1900).

99   Annett, H. E., Dutton, J. E., & Elliott, J. H. *Report of the Malaria Expedition to Nigeria of the Liverpool School of Tropical Medicine and Medical Parasitology.* (University Press of Liverpool, 1901).

100 Ross, R. The malaria expedition to Sierra Leone. *British Medical Journal* **September 9, 16, 30; October 14** (1899).

101 Ross, R., Annett, H. E., & Austen, E. E. *Report of the Malaria Expedition of the Liverpool School of Tropical Medicine and Medical Parasitology.* (University Press of Liverpool, 1900).

102 Dunlap, T. R. *DDT, Silent Spring, and the Rise of Environmentalism.* (University of Washington Press, 2008).

103 Rush, B. *An Account of the Bilious Remitting Yellow Fever, as it Appeared in the City of Philadelphia, in the Year 1793*. (Thomas Dobson, 1794).

104 Carter, H. R. *Yellow Fever: An Epidemiological and Historical Study of Its Place of Origin*. (Williams & Wilkins Co., 1931).

105 Creighton, C. The origin of yellow fever. *North American Review* **139**, 335–47 (1884).

106 Murphy, J. *An American Plague*. (Clarion Books, 2003).

107 Carey, M. *A Short Account of the Malignant Fever, Lately Prevalent in Philadelphia: with a Statement of the Proceedings that Took Place on the Subject, in Different Parts of the United States*. (Mathew Carey, 1793).

108 Jones, A., & Allen, R. *A Narrative of the Proceedings of the Black People, During the Late Awful Calamity in Philadelphia, in the Year 1793: and a Refutation of Some Censures, Thrown upon Them in Some Late Publications*. (William W. Woodward, 1794).

109 Otter, S. *Philadelphia Stories: America's Literature of Race and Freedom*. (Oxford University Press, 2010).

110 Washington, G. To James Madison from George Washington. **October 14, 1793**. Founders Online, National Archives. (1793). https://founders.archives .gov/documents/Madison/01-15-02-0081.

111 *Minutes of the Proceedings of the Committee, Appointed on the 14th September, 1793, by the Citizens of Philadelphia, the Northern Liberties, and the District of Southwark, to Attend to and Alleviate the Sufferings of the Afflicted with the Malignant Fever, Prevalent in the City and its Vicinity*. (City of Philadelphia, 1848).

112 Jefferson, T. Letter to Benjamin Rush. **September 23, 1800**. Founders Online, National Archives. (1800). https://founders.archives.gov/documents /Jefferson/01-32-02-0102.

113 Adams, J. Letter to Thomas Jefferson. **June 30, 1813**. Founders Online, National Archives. (1813). https://founders.archives.gov/documents /Jefferson/03-06-02-0216.

114 Stapleton, D. H., & Carter, E. C. I. "I have the itch of botany, of chemistry, of mathematics . . . strong upon me": the science of Benjamin Henry Latrobe. *Proceedings of the American Philosophical Society* **128**, 173–92 (1984).

115 Sherman, I. W. *Twelve Diseases That Changed Our World*. (ASM Press, 2007).

116 Choppin, S. History of the importation of yellow fever into the United States, from 1693–1878. *Public Health Papers, American Public Health Association* **4**, 190–206 (1878).

117 The burning of the quarantine hospital on Staten Island. *Harper's Weekly* **September 11** (1858).

118 Message from the president of the United States, transmitting certain papers in regard to experiments conducted for the purpose of coping with yellow fever. Senate Document No. 10, 59th Congress, 2d Session (1907).

119 Faust, E. C. History of human parasitic infections. *Public Health Reports* **70**, 958–65 (1955).

120 Souchon, E. Educational points concerning yellow fever, to be spread broadcast by the press, pulpit, school-teachers and others, and by all men of good will. Louisiana State Board of Health (1898).

121 The yellow fever plot. *New York Times* **May 16**, 4 (1865).

122 Segel, L. "The yellow fever plot": germ warfare during the Civil War. *Canadian Journal of Diagnosis* **September**, 47–50 (2002).

123 Quinn, D. A. *Heroes and Heroines of Memphis, or Reminiscences of the Yellow Fever Epidemics that Afflicted the City of Memphis During the Autumn Months of 1873, 1878, and 1879, to Which is Added a Graphic Description of Missionary Life in Eastern Arkansas*. (E. L. Freeman & Son, 1887).

124 Ffirth, S. *A Treatise on Malignant Fever; with an Attempt to Prove its Non-contagious Nature*. (B. Graves, 1804).

125 Michel, R. F. Epidemic of yellow fever in Montgomery, Alabama, summer of 1873. *Transactions of the Medical Association of the State of Alabama* **1874**, 84–111 (1874).

126 Dromgoole, J. P. *Dr. Dromgoole's Yellow Fever Heroes, Honors, and Horrors of 1878*. (John P. Morton & Co., 1879).

127 Mitchell, J. Account of the yellow fever which prevailed in Virginia in the years 1737, 1741, and 1742, in a letter to the late Cadwallader Colden, Esq. of New-York. *American Medical and Philosophical Register* **4**, 181–215 (1814).

128 Rush, B. Letter to Julia Rush. October 27, 1793. In *Letters of Benjamin Rush, Volume 2: 1793–1813* (ed. L. H. Butterfield). (Princeton University Press, 1951).

129 Holt, J. Analysis of the records of yellow fever in New Orleans, in 1876. *New Orleans Medical and Surgical Association* **November 11** (1876).

130 Erskine, J. H. A report on yellow fever as it appeared in Memphis, Tenn., in 1873. *Public Health Papers and Reports* **1**, 385–92 (1873).

131 *Conclusions of the Board of Experts Authorized by Congress to Investigate the Yellow Fever Epidemic of 1878*. (Judd & Detweiler, 1879).

132 Agramonte, A. The inside history of a great medical discovery. *Scientific Monthly* **1**, 209–37 (1915).

133 Reed, W. Letter from Walter Reed to Emilie Lawrence Reed. **December 31, 1900**. Philip S. Hench Walter Reed Yellow Fever Collection, University of Virginia.

134 Nott, J. C. The cause of yellow fever. *New Orleans Medical and Surgical Journal* **4**, 563–601 (1848).

135 Agramonte, A. An account of Dr. Louis-Daniel Beauperthuy, a pioneer in yellow fever research. *Boston Medical and Surgical Journal* **June 18**, 927–30 (1908).

136 Finlay, C. The mosquito hypothetically considered as an agent in the trans- mission of yellow fever poison. *New Orleans Medical and Surgical Journal* **1881–82**, 601–16 (1882).

137 Reed, W., Carroll, J., Agramonte, A., & Lazear, J. W. The etiology of yellow fever: a preliminary note. *Philadelphia Medical Journal* **October 27**, 37–53 (1900).

138 Finlay, C. *Selected Papers of Dr. Carlos J. Finlay.* (Republica de Cuba, Secretaria de Sanidad y Beneficencia, 1912).

139 Sternberg, G. M. The transmission of yellow fever by mosquitoes. *Popular Science Monthly* **59** (1901).

140 Kelly, H. A. *Walter Reed and Yellow Fever.* (McClure, Phillips & Co., 1906).

141 Smith, T., & Kilborne, F. L. Investigations into the nature, causation, and prevention of southern cattle fever. In *Bureau of Animal Industry, Eighth and Ninth Annual Reports for the Years 1891–1892*, 177–304. (US Government Printing Office, 1893).

142 Bruce, D. *Preliminary Report on the Tsetse Fly Disease or Nagana.* (Bennett & Davis, 1895).

143 Crosby, M. C. *The American Plague.* (Berkley Books, 2006).

144 Sanarelli, G. A lecture on yellow fever, with a description of the *Bacillus icter- oides*. *British Medical Journal* **July 3**, 7–11 (1897).

145 Carroll, J. A brief review of the aetiology of yellow fever. *New York Medi- cal Journal and Philadelphia Medical Journal, Consolidated* **February 6, 13** (1904).

146 Reed, W., & Carroll, J. A comparative study of the biological characters and pathogenesis of *Baccillus* X (Sternberg), *Baccillus icteroides* (Sanarelli), and the hog-cholera *Bacillus* (Salmon and Smith). *Journal of Experimental Medi- cine* **5**, 215–70 (1900).

147 Reed, W. Recent researches concerning the etiology, propagation, and pre- vention of yellow fever, by the United States Army Commission. *Journal of Hygiene* **2**, 101–19 (1902).

148 Craig, S. C. *In the Interest of Truth: The Life and Science of Surgeon General George Miller Sternberg.* (Office of the Surgeon General, Borden Institute, 2013).

149 Petri, W. A. J. America in the world: 100 years of tropical medicine and hygiene. *American Journal of Tropical Medicine and Hygiene* **71**, 2–16 (2004).

150 Lazear, M. H. Letter from Mabel Houston Lazear to James Carroll. **Novem- ber 10, 1900.** Philip S. Hench Walter Reed Yellow Fever Collection, Univer- sity of Virginia.

151 Reed, W., Carroll, J., & Agramonte, A. The etiology of yellow fever: an ad- ditional note. *Journal of the American Medical Association* **36**, 431–40 (1901).

152 Moran, J. J. Memoirs of a human guinea pig. Philip S. Hench Walter Reed Yellow Fever Collection, University of Virginia (1948).

153 Carey, F. 50 years ago Reed faced "yellow jack" in Havana. Associated Press **June 25** (1950).

154 Reed, W. *The Propagation of Yellow Fever—Observations Based on Recent Researches*. (US Government Printing Office, 1911).

155 Reed, W., Carroll, J., & Agramonte, A. Experimental yellow fever. *American Medicine* **2**, 15–23 (1901).

156 Finlay, C. E. Dr. Carlos J. Finlay's positive cases of experimental yellow fever. *New Orleans Medical and Surgical Journal* **69**, 333–43 (1917).

157 Agramonte, A. Finlay and Delgado's experimental yellow fever (a reply to Dr. C. E. Finlay). *New Orleans Medical and Surgical Journal* **69**, 344–51 (1917).

158 Guiteras, J. Experimental yellow fever at the inoculation station of the Sanitary Department of Havana with a view to producing immunization. *American Medicine* **3**, 809–17 (1901).

159 Reed, W., & Carroll, J. The etiology of yellow fever: a supplemental note. *American Medicine* **February 22**, 301–5 (1902).

160 Adams, C. F. The Panama Canal Zone: an epochal event in sanitation. *Proceedings of the Massachusetts Historical Society* **17**, 1–38 (1911).

161 Halstead, M. *The Illustrious Life of William McKinley our Martyred President*. (By the author, 1901).

162 *Discussion of the Paper of Drs. Reed and Gorgas*. (Berlin Printing Company, 1902).

163 One of McKinley's surgeons passes away. *Hawaiian Star* **December 6**, 2 (1911).

164 Gorgas, W. C. *A Few General Directions with Regard to Destroying Mosquitoes, Particularly the Yellow Fever Mosquito*. (US Government Printing Office, 1904).

165 Gorgas, W. C. *Sanitation in Panama*. (Appleton, 1915).

166 Gorgas, M. D., & Hendricks, B. J. *William Crawford Gorgas: His Life and Work*. (Doubleday, Page & Co., 1924).

167 *The Rockefeller Foundation Annual Report 1926*. (Rockefeller Foundation, 1926).

168 Stokes, A., Bauer, J. H., & Hudson, N. P. Experimental transmission of yellow fever virus to laboratory animals. *American Journal of Tropical Medicine* **8**, 103–64 (1928).

169 Bryan, C. S. Discovery of the yellow fever virus. *International Journal of Infectious Diseases* **2**, 52–54 (1997).

170 *The Rockefeller Foundation Annual Report 1927*. (Rockefeller Foundation, 1927).

171 Hudson, N. P. Adrian Stokes and yellow fever research: a tribute. *Transactions of the Royal Society of Tropical Medicine and Hygiene* **60**, 170–74 (1966).

172 Porterfield, J. S. Yellow fever in west Africa: a retrospective glance. *British Medical Journal* **299**, 1555–57 (1989).

173 Berry, G. P., & Kitchen, S. F. Yellow fever accidentally contracted in the laboratory. *American Journal of Tropical Medicine* **11**, 365–434 (1931).

174 Bauer, J. H. Transmission of yellow fever by mosquitoes other than *Aedes aegypti*. *American Journal of Tropical Medicine* **8**, 261–82 (1928).

175 Delatte, H., et al. The invaders: phylogeography of dengue and chikungunya virus *Aedes* vectors, on the south west islands of the Indian Ocean. *Infection, Genetics and Evolution* **11**, 1769–81 (2011).

176 Pialoux, G., Gaüzère, B. A., Jauréguiberry, S., & Strobel, M. Chikungunya, an epidemic arbovirosis. *Lancet Infectious Diseases* **7**, 319–27 (2007).

177 Bergstrand, H. *The Nobel Prize in Physiology or Medicine 1951, Award Ceremony Speech*. (Elsevier Publishing Co., 1951).

178 Theiler, M. Susceptibility of white mice to the virus of yellow fever. *Science* **71**, 367 (1930).

179 Theiler, M. Studies on the action of yellow fever virus on mice. *Annals of Tropical Medicine & Parasitology* **24**, 249–72 (1930).

180 Smith, H. H. Yellow fever vaccination with cultured virus (17D) without immune serum. *American Journal of Tropical Medicine and Hygiene* **18**, 437–68 (1938).

181 Theiler, M., & Smith, H. H. The effect of prolonged cultivation in vitro upon the pathogenicity of yellow fever virus. *Journal of Experimental Medicine* **65**, 767–86 (1937).

182 Frierson, J. G. The yellow fever vaccine: a history. *Yale Journal of Biology and Medicine* **83**, 77–85 (2010).

183 Cirillo, V. J. Two faces of death: fatalities from disease and combat in America's principal wars, 1775 to present. *Perspectives in Biology and Medicine* **51**, 121–33 (2008).

184 Peltier, M. Vaccination mixte contre la fièvre jaune et la variole sur des populations indigènes du Sénégal. *Annales de l Institut Pasteur (Dakar)* **65**, 146–69 (1940).

185 Durieux, C. Mass yellow fever vaccination in French Africa south of the Sahara. In *Yellow Fever Vaccination* (ed. K. Smithburn), 115–21. (World Health Organization, 1956).

186 Norrby, E. Yellow fever and Max Theiler: the only Nobel Prize for a virus vaccine. *Journal of Experimental Medicine* **204**, 2779–84 (2007).

187 Mathis, C., Sellards, A. W., & Laigret, J. Sensibilité du *Macacus rhesus* au virus fièvre jaune. *Comptes rendus de l'Académie des Sciences* **186**, 604–6 (1928).

188 Rice, C. M. Nucleotide sequence of yellow fever virus: implications for flavivirus gene expression and evolution. *Science* **229**, 726–33 (1985).

189 Snyder, J. C. The typhus fevers. In *Viral and Rickettsial Infections of Man* (ed. T. M. Rivers & F. L. Horsfall). (J. B. Lippincott Co., 1959).

190 Howard, J. *The State of the Prisons in England and Wales*. (Warrington, 1777).

191 *Encyclopedia of Plague and Pestilence from Ancient Times to the Present*. (Facts on File, 2008).

192 Ackerknecht, E. H. *History and Geography of the Most Important Diseases*. (Hafner Publishing Co., Inc., 1965).

193 Cartwright, F. F., & Biddiss, M. *Disease & History*, 2d ed. (Sutton Publishing, 2004).

194 Schultz, M. G., & Morens, D. M. Charles-Jules-Henri Nicolle. *Emerging Infectious Diseases* **15**, 1520–22 (2009).

195 *Nobel Lectures, Physiology or Medicine, 1922–1941*. (Elsevier Publishing Co., 1965).

196 Gross, L. How Charles Nicolle of the Pasteur Institute discovered that epidemic typhus is transmitted by lice: reminiscences from my years at the Pasteur Institute in Paris. *Proceedings of the National Academy of Sciences USA* **93**, 10539–40 (1996).

197 Nicolle, C., Comte, C., & Conseil, E. Transmission expérimentale du typhus exanthématique par le pou du corps. *Comptes-rendus hebdomadaires des séances de l'Académie des Sciences* **149**, 486–89 (1909).

198 Ricketts, H. T., & Wilder, R. M. The transmission of the typhus fever of Mexico (Tabardillo) by means of the louse (*Pediculus vestimenti*). *Journal of the American Medical Association* **54**, 1304–7 (1910).

199 Da Rocha-Lima, H. Zur aetiologie des fleckfiebers. *Berliner Klinische Wochenschrift* **53**, 567–69 (1916).

200 Von Prowazek, S. Ätiologische Untersuchungen über den Flecktyphus in Serbien 1913 und Hamburg 1914. *Beitrage zur Klinik der Infektionskrankheiten und zur Immunitätsforschung* **4**, 5–31 (1914).

201 Paape, H. Imprisonment and deportation. In *The Diary of Anne Frank: The Critical Edition* (Doubleday, 1986).

202 Zinsser, H. Varieties of typhus virus and the epidemiology of the American form of European typhus fever (Brill's disease). *American Journal of Hygiene* **20**, 513–32 (1934).

203 *The Jerusalem Bible*. (Koren Publishers, 1983).

204 Rosen, W. *Justinian's Flea—Plague, Empire, and the Birth of Europe*. (Viking, 2007).

205 Mommsen, T. E. Petrarch's conception of the "Dark Ages." *Speculum* **17**, 226–42 (1942).

206 Kitasato, S., & Nakagawa, A. Plague. In *Twentieth Century Practice: An International Encyclopedia of Modern Medical Science by Leading Authorities of Europe and America, Vol. 15: Infectious Diseases* (ed. T. L. Stedman). (William Wood & Co., 1898).

207 Aberth, J., ed. *The Black Death: The Great Mortality of 1348–1350: A Brief History with Documents*. (Bedford/St. Martin's, 2005).

208 Gregoras, N. *Byzantine History*. (1359). In *The Black Death: The Great Mortality of 1348–1350: A Brief History with Documents* (ed. J. Aberth). (Bedford/St. Martin's, 2005).

209 Derbes, V. De Mussis and the Great Plague of 1348: a forgotten episode of bacteriological warfare. *Journal of the American Medical Association* **196**, 59–62 (1966).

210 Ibn al-Wardī, A. H. U. *Essay on the Report of the Pestilence*. (1348). In *The Black Death: The Great Mortality of 1348–1350: A Brief History with Documents* (ed. J. Aberth). (Bedford/St. Martin's, 2005).

211 Petrarch, F. *Letters on Familiar Matters*. (1349). In *The Black Death: The Great Mortality of 1348–1350: A Brief History with Documents* (ed. J. Aberth). (Bedford/St. Martin's, 2005).

212 Boccaccio, G. *The Decameron*. (1349–51). In *The Black Death: The Great Mortality of 1348–1350: A Brief History with Documents* (ed. J. Aberth). (Bedford/St. Martin's, 2005).

213 D'Agramont, J. *Regimen of Protection against Epidemics*. (1348). In *The Black Death: The Great Mortality of 1348–1350: A Brief History with Documents* (ed. J. Aberth). (Bedford/St. Martin's, 2005).

214 Pedro IV of Aragon. *Response to Jewish Pogrom of Tárrega*. (1349). In *The Black Death: The Great Mortality of 1348–1350: A Brief History with Documents* (ed. J. Aberth). (Bedford/St. Martin's, 2005).

215 *Takkanoth (Accord) of Barcelona*. (1354). In *The Black Death: The Great Mortality of 1348–1350: A Brief History with Documents* (ed. J. Aberth). (Bedford/St. Martin's, 2005).

216 *Interrogation of the Jews of Savoy*. (1348). In *The Black Death: The Great Mortality of 1348–1350: A Brief History with Documents* (ed. J. Aberth). (Bedford/St. Martin's, 2005).

217 Mathias of Neuenburg. *Chronicle*. (1349–50). In *The Black Death: The Great Mortality of 1348–1350: A Brief History with Documents* (ed. J. Aberth). (Bedford/St. Martin's, 2005).

218 Konrad of Megenberg. *Concerning the Mortality in Germany*. (1350). In *The Black Death: The Great Mortality of 1348–1350: A Brief History with Documents* (ed. J. Aberth). (Bedford/St. Martin's, 2005).

219 Pope Clement VI. *Sicut Judeis (Mandate to Protect the Jews)*. (1348). In *The Black Death: The Great Mortality of 1348–1350: A Brief History with Documents* (ed. J. Aberth). (Bedford/St. Martin's, 2005).

220 Closener, F. *Chronicle*. (1360–62). In *The Black Death: The Great Mortality of 1348–1350: A Brief History with Documents* (ed. J. Aberth). (Bedford/St. Martin's, 2005).

221 Medical Faculty of the University of Paris. *Consultation*. (1348). In *The Black Death: The Great Mortality of 1348–1350: A Brief History with Documents* (ed. J. Aberth). (Bedford/St. Martin's, 2005).

222 Sanctus, L. Letter. (1348). In *The Black Death: The Great Mortality of 1348–1350: A Brief History with Documents* (ed. J. Aberth). (Bedford/St. Martin's, 2005).

223 Villani, G. *Chronicle*. (1348). In *The Black Death: The Great Mortality of 1348–1350: A Brief History with Documents* (ed. J. Aberth). (Bedford/St. Martin's, 2005).

224 Di Tura, A. *Sienese Chronicle*. (1348–51). In *The Black Death: The Great Mortality of 1348–1350: A Brief History with Documents* (ed. J. Aberth). (Bedford/St. Martin's, 2005).

225 Ibn al-Khatīb, L. A. I. *A Very Useful Inquiry into the Horrible Sickness*. (1349–52). In *The Black Death: The Great Mortality of 1348–1350: A Brief History with Documents* (ed. J. Aberth). (Bedford/St. Martin's, 2005).

226 Da Foligno, G. *Short Casebook*. (1348). In *The Black Death: The Great Mortality of 1348–1350: A Brief History with Documents* (ed. J. Aberth). (Bedford/St. Martin's, 2005).

227 Ibn Khātima, A. J. A. *Description and Remedy for Escaping the Plague*. (1349). In *The Black Death: The Great Mortality of 1348–1350: A Brief History with Documents* (ed. J. Aberth). (Bedford/St. Martin's, 2005).

228 Shakespeare, W. *The Most Excellent and Lamentable Tragedie of Romeo and Juliet*. (Thomas Creede & Cuthbert Burby, 1599). In *The Works of William Shakespeare, the Text Revised by the Rev. Alexander Dyce, In Ten Volumes*, vol. 1. 4th ed. London: Bickers & Son, 1880.

229 Liston, W. G. Plague, rats and fleas. *Journal of the Bombay Natural History Society* **16**, 253–74 (1905).

230 Cantlie, J. The plague in Hong Kong. *British Medical Journal* **2**, 423–27 (1894).

231 The plague at Hong Kong. *British Medical Journal* **2**, 201 (1894).

232 The plague at Hong Kong. *Lancet* **2**, 269–70 (1894).

233 Solomon, T. Hong Kong, 1894: the role of James A. Lowson in the controversial discovery of the plague bacillus. *Lancet* **350**, 59–62 (1997).

234 Lagrange, E., Liège, M. D., & Paris, D. T. M. Concerning the discovery of the plague bacillus. *Journal of Tropical Medicine and Hygiene* **29**, 299–303 (1926).

235 Plague in the Far East. *British Medical Journal* **August 22**, 460 (1896).

236 The plague in Hong-Kong in 1894: a story of Chinese antipathies. *Lancet* **April 4**, 936 (1896).

237 Lowson, J. A. *The Epidemic of Bubonic Plague in Hong Kong 1894*. (Government Printer, 1895).

238 Lee, P.-T. Colonialism versus nationalism: the plague of Hong Kong in 1894. *Journal of Northeast Asian History* **10**, 97–128 (2013).

239 Obituary: Baron Shibasaburo Kitasato. *British Medical Journal* **June 27,** 1141–42 (1931).

240 Kitasato, S. The bacillus of bubonic plague. *Lancet* **2**, 428–30 (1894).

241 Gross, L. How the plague bacillus and its transmission through fleas were discovered: reminiscences from my years at the Pasteur Institute in Paris. *Proceedings of the National Academy of Sciences USA* **92**, 7609–11 (1995).

242 Hawgood, B. J. Alexandre Yersin (1863–1943): discoverer of the plague bacillus, explorer and agronomist. *Journal of Medical Biogeography* **16**, 167–72 (2008).

243 Schwartz, M. The Institut Pasteur: 120 years of research in microbiology. *Research in Microbiology* **159**, 5–14 (2008).

244 Kousoulis, A. A., Karamanou, M., Tsoucalas, G., Dimitriou, T., & Androutsos, G. Alexandre Yersin's explorations (1892–1894) in French Indochina before the discovery of the plague bacillus. *Acta Medico-Historica Adriatica* **10**, 303–10 (2012).

245 The plague at Hong Kong. *Lancet* **1**, 1581–82 (1894).

246 Yersin, A. Le peste bubonique à Hong-Kong. *Annales de l'Institut Pasteur* **8**, 662–67 (1894).

247 Crawford, E. A. J. Paul-Louis Simond and his work on plague. *Perspectives in Biology and Medicine* **39**, 446–58 (1996).

248 The plague in China. *Lancet* **August 4**, 266 (1894).

249 The plague at Hong Kong. *Lancet* **2**, 325 (1894).

250 The bacillus of plague. *British Medical Journal* **2**, 369–70 (1894).

251 The plague in Hong-Kong. *Lancet* **2**, 391–92 (1894).

252 Yabe, T. The microbe of plague. *Journal of Tropical Medicine* **4**, 59–60 (1901).

253 The late Baron Shibasaburo Kitasato. *Canadian Medical Association Journal* **August**, 206 (1931).

254 Millott Severn, A. G. A note concerning the discovery of the *Bacillus pestis*. *Journal of Tropical Medicine and Hygiene* **August 15**, 208–9 (1927).

255 Biographical sketch: Alexandre Yersin (1863–1943). Pasteur Institute Archives and Collection. http://www.pasteur.fr/infosci/archives/e_yer0.html.

256 Hawgood, B. J. Alexandre Yersin MD (1863–1943); Suoi Dau near Nha Trang, Vietnam. *Journal of Medical Biogeography* **19**, 138 (2011).

257 Simond, P.-L. La propagation de la peste. *Annales de l'Institut Pasteur* **12**, 625–87 (1898).

258 Köhler, W., & Köhler, M. Plague and rats, the "Plague of the Philistines," and: what did our ancestors know about the role of rats in plague. *International Journal of Medical Microbiology* **293**, 333–40 (2003).

259 Lowson, J. A. The bacteriology of plague. *British Medical Journal* **January 23**, 237–38 (1897).

260 Reports on plague investigations in India. *Journal of Hygiene* **6**, 421–536 (1906).

261 Rennie, A. The plague in the East. *British Medical Journal* **September 15,** 615–16 (1894).

262 Low, B. Report upon the progress and diffusion of bubonic plague from 1879–1898. In *Twenty-eighth Annual Report of the Local Government Board 1898–1899. Supplement Containing the Report of the Medical Officer for 1898–1899.* (Darling & Son, Ltd., 1899).

263 Ogata, M. Ueber die Pestepidemie in Formosa. *Centralblatt für Bakteriologie und Parasitenkunde* **21,** 774 (1897).

264 Biographical sketch: Paul-Louis Simond (1858–1947). Pasteur Institute Archives and Collection. http://www.pasteur.fr/infosci/archives/e_sim0.html.

265 Simond, M., Godley, M. L., & Mouriquand, P. D. E. Paul-Louis Simond and his discovery of plague transmission by rat fleas: a centenary. *Journal of the Royal Society of Medicine* **91,** 101–4 (1998).

266 Simond, P.-L. Comment fut mis en évidence le rôle de la puce dans la transmission de la peste. *Revue d'hygiène* **58,** 5–17 (1936).

267 Gauthier, J. O., & Raybaud, A. Recherches expérimentales sur le rôle des parasites du rat dans la transmission de la peste. *Revue d'hygiène* **25,** 426–38 (1903).

268 Löwy, I., & Rodhain, F. Paul-Louis Simond and yellow fever. *Bulletin de la Société de Pathologie Exotique* **92,** 392–95 (1999).

269 Bacot, A. W., & Martin, C. J. Observations on the mechanism of the transmission of plague by fleas. *Journal of Hygiene* **13 (Plague supplement 3),** 423–39 (1914).

270 Obituary. Arthur Bacot. *Nature* **109,** 618–20 (1922).

271 Chouikha, I., & Hinnebusch, B. J. *Yersinia*-flea interactions and the evolution of the arthropod-borne transmission route of plague. *Current Opinion in Microbiology* **15,** 239–46 (2012).

272 Bacot, A. W. A study of the bionomics of the common rat fleas and other species associated with human habitations, with special reference to the influence of temperature and humidity at various periods of the life history of the insect. *Journal of Hygiene* **13 (Plague supplement 3),** 447–653 (1914).

273 Perry, R. D., & Fetherson, J. D. *Yersinia pestis*-etiologic agent of plague. *Clinical Microbiology Reviews* **10,** 35–66 (1997).

274 Raoult, D., et al. Molecular identification by "suicide PCR" of *Yersinia pestis* as the agent of Medieval Black Death. *Proceedings of the National Academy of Sciences USA* **97,** 12800–803 (2000).

275 Haensch, S., et al. Distinct clones of *Yersinia pestis* caused the black death. *PLoS Pathogens* **6,** e1001134 (2010).

276 Harbeck, M., et al. *Yersinia pestis* DNA from skeletal remains from the 6th century AD reveals insights into Justinianic Plague. *PLoS Pathogens* **9,** e1003349 (2013).

277 Achtman, M., et al. Microevolution and history of the plague bacillus, *Yersinia pestis*. *Proceedings of the National Academy of Sciences USA* **101**, 17837–42 (2004).

278 Bos, K. I., et al. A draft genome of *Yersinia pestis* from victims of the Black Death. *Nature* **478**, 506–10 (2011).

279 Bos, K. I., et al. Eighteenth-century *Yersinia pestis* genomes reveal the long-term persistence of an historical plague focus. *eLife* **5**, e12994 (2016).

280 Wagner, D. M., et al. *Yersinia pestis* and the Plague of Justinian 541–43 AD: a genomic analysis. *Lancet Infectious Diseases* **14**, 319–26 (2014).

281 Montenegro, J. V. *Bubonic Plague: Its Course and Symptoms and Means of Prevention and Treatment*. (William Wood & Co., 1900).

282 Kupferschmidt, H. History of the epidemiology of plague: changes in the understanding of plague epidemiology since the discovery of the plague pathogen in 1894. *Antimicrobics and Infectious Diseases Newsletter* **16**, 51–53 (1997).

283 Doriga, Dr. The prevention of plague through the suppression of rats and mice. *Public Health* **12**, 92–98 (1899).

284 Pelletier, P. J., & Caventou, J. B. Note sur un nouvel alkalai. *Annales de chimie et de physique* **8**, 323–24 (1818).

285 Bacot, A. W. The effect of the vapours of various insecticides upon fleas (*Ceratophyllus fasciatus* and *Xenopsylla cheopis*) at each stage in their life-history and upon the bed bug (*Cimex lectularius*) in its larval stage. *Journal of Hygiene* **13** (**Plague supplement 3**), 665–81 (1914).

286 Runge, F. F. Ueber einige Produkte der Steinkohlendestillation. *Annalen der Physik und Chemie* **31**, 65–78 (1834).

287 *Trials of War Criminals before the Nuernberg Military Tribunals, Vol. 1: The Medical Case*. (US Government Printing Office, 1946–49).

288 Richardson, B. W. Greek fire: its ancient and modern history. *Popular Science Review* **3**, 164–77 (1864).

289 Biographical note, Thucydides, c. 460–c. 400 B.C. In *Great Books of the Western World, Vol. 6: The History of Herodotus and The History of the Peloponnesian War of Thucydides* (ed. R. M. Hutchins). (William Benton; Encyclopaedia Britannica, Inc., 1952).

290 Thucydides. *The History of the Peloponnesian War* (trans. R. Crawley; rev. R. Feetham). In *Great Books of the Western World, Vol. 6: The History of Herodotus and The History of the Peloponnesian War of Thucydides* (ed. R. M. Hutchins). (William Benton; Encyclopaedia Britannica, Inc., 1952).

291 Joy, R. J. T. Historical aspects of medical defense against chemical warfare. In *Medical Aspects of Chemical and Biological Warfare* (ed. F. R. Sidell, E. T. Takafuji & D. R. Franz). (Office of the Surgeon General at TMM Publications, 1997).

292 Gibbon, E. *The History of the Decline and Fall of the Roman Empire, Volume the Tenth*. (Luke White, 1788).

293 Lloyd, C. *Lord Cochrane: Seaman—Radical—Liberator*. (Longmans, Green & Co., 1947).

294 *The Panmure Papers*. Vol. 1. (Hodder & Stoughton, 1908).

295 Mendelssohn, K. *The World of Walther Nernst: The Rise and Fall of German Science 1864–1941*. (University of Pittsburgh Press, 1973).

296 Reid, W. *Memoirs and Correspondence of Lyon Playfair*. (Cassell & Co., Ltd., 1899).

297 Fries, A. A., & West, C. J. *Chemical Warfare*. (McGraw-Hill Book Co., Inc., 1921).

298 Waitt, A. H. *Gas Warfare: The Chemical Weapon, Its Use, and Protection against It*. (Duell, Sloan & Pearce, 1942).

299 Miles, W. D. The idea of chemical warfare in modern times. *Journal of the History of Ideas* **31**, 297–304 (1970).

300 Wöhler, F. Ueber künstliche Bildung des Harstoffs. *Annalen der Physik und Chemie* **88**, 253–56 (1828).

301 Goran, M. *The Story of Fritz Haber*. (University of Oklahoma Press, 1967).

302 Joy, C. A. Biographical sketch of Frederick Wöhler. *Popular Science Monthly* **17** (1880).

303 Russell, E. *War and Nature*. (Cambridge University Press, 2001).

304 *Nobel Lectures, Chemistry, 1922–1941*. (Elsevier Publishing Co., 1966).

305 Willstätter, R. *From My Life: The Memoirs of Richard Willstätter* (trans. L. Hornig). (Verlag Chemie, GmbH, 1949).

306 Renn, J. Introduction. In *One Hundred Years of Chemical Warfare: Research, Deployment, Consequences* (ed. B. Friedrich et al.). (SpringerOpen, 2017).

307 Ertle, G. Fritz Haber and his institute. In *One Hundred Years of Chemical Warfare: Research, Deployment, Consequences* (ed. B. Friedrich et al.). (SpringerOpen, 2017).

308 Lemmerich, J. *Science and Conscience: The Life of James Franck*. (Stanford University Press, 2011).

309 Friedrich, B., & Hoffmann, D. Clara Immerwahr: a life in the shadow of Fritz Haber. In *One Hundred Years of Chemical Warfare: Research, Deployment, Consequences* (ed. B. Friedrich et al.). (SpringerOpen, 2017).

310 Friedrich, B., & James, J. From Berlin-Dahlem to the fronts of World War I: the role of Fritz Haber and his Kaiser Wilhelm Institute in German chemical warfare. In *One Hundred Years of Chemical Warfare: Research, Deployment, Consequences* (ed. B. Friedrich et al.). (SpringerOpen, 2017).

311 Hill, B. A. History of the medical management of chemical casualties. In *Medical Aspects of Chemical Warfare* (ed. S. D. Tuorinsky). (Borden Institute, Walter Reed Army Medical Center, 2008).

312 Szöllösi-Janze, M. The scientist as expert: Fritz Haber and German chemical warfare during the First World War and beyond. In *One Hundred Years of Chemical Warfare: Research, Deployment, Consequences* (ed. B. Friedrich et al.). (SpringerOpen, 2017).

313 von Hippel, F. James Franck: science and conscience. *Physics Today* **June 2010**, 41–46 (2010).

314 Rice, S. A., & Jortner, J. *James Franck 1882–1964*. (National Academy of Sciences, 2010).

315 Smart, J. K. History of chemical and biological warfare: an American perspective. In *Medical Aspects of Chemical and Biological Warfare* (ed. F. R. Sidell, E. T. Takafuji & D. R. Franz) (Office of the Surgeon General at TMM Publications, 1997).

316 Haber, F. Chemistry in war (a translation of *Fünf Vorträge, aus den Jahren 1920–1923*). *Journal of Chemical Education* **November**, 526–29, 553 (1945).

317 Carter, C. F. Growth of the chemical industry. *Current History* **15**, 423–28 (1922).

318 Baker, N. D. Chemistry in warfare. *Journal of Industrial and Engineering Chemistry* **11**, 921–23 (1919).

319 Higgs, R. The boll weevil, the cotton economy, and black migration 1910–1930. *Agricultural History* **50**, 335–50 (1976).

320 Dunlap, T. R. *DDT: Scientists, Citizens, and Public Policy*. (Princeton University Press, 1981).

321 Howard, L. O., & Popenoe, C. H. Hydrocyanic-acid gas against household insects. *US Department of Agriculture, Bureau of Entomology Circular* **163**, 1–8 (1912).

322 Howard, L. O. Entomology and the war. *Scientific Monthly* **8**, 109–17 (1919).

323 Had deadliest gas ready for Germans; "Lewisite" might have killed millions. *New York Times* **May 25** (1919).

324 The fly must be exterminated to make the world safe for habitation. *American City* **19**, 12 (1918).

325 Broadberry, S., & Harrison, M. The economics of World War I: an overview. In *The Economics of World War I* (ed. S. Broadberry & M. Harrison). (Cambridge University Press, 2005).

326 Borkin, J. *The Crime and Punishment of I. G. Farben*. (Free Press, 1978).

327 Churchill, W. *Thoughts and Adventures*. (Odhams Press Ltd., 1932).

328 Man versus insects: the next great war. *Advertiser* **August 21**, 18 (1915).

329 Forbes, S. A. *The insect, the farmer, the teacher, the citizen, and the state*. (Illinois State Laboratory of Natural History, 1915).

330 Arrhenius, S. On the influence of carbonic acid in the air upon the temperature of the ground. *London, Edinburgh, and Dublin Philosophical Magazine and Journal of Science* **5**, 237–76 (1896).

331 *Nobel Lectures, Chemistry, 1901–1921.* (Elsevier Publishing Co., 1966).

332 Arrhenius, G., Caldwell, K., & Wold, S. A tribute to the memory of Svante Arrhenius (1859–1927). *Annual Meeting of the Royal Swedish Academy of Engineering Sciences* (2008).

333 Weindling, P. The uses and abuses of biological technologies: Zyklon B and gas disinfestation between the First World War and the Holocaust. *History and Technology* **11**, 291–98 (1994).

334 Stoltzenberg, D. *Fritz Haber: Chemist, Nobel Laureate, German, Jew.* (Chemical Heritage Press, 2004).

335 *The Treaty of Peace between the Allied and Associated Powers and Germany, the Protocol Annexed Thereto, the Agreement respecting the Military Occupation of the Territories of the Rhine, and the Treaty between France and Great Britain respecting Assistance to France in the Event of Unprovoked Aggression by Germany. Signed at Versailles, June 28th, 1919.* (His Majesty's Stationery Office, 1919).

336 *Nobel Lectures, Physics, 1901–1921.* (Elsevier Publishing Co., 1967).

337 Born, M. Arnold Johannes Wilhelm Sommerfeld, 1868–1951. *Obituary Notices of Fellows of the Royal Society* **8**, 274–96 (1952).

338 *Nobel Lectures, Physics, 1922–1941.* (Elsevier Publishing Co., 1965).

339 The Nobel medals and the medal for the prize in economic sciences. http://www.nobelprize.org/nobel_prizes/about/medals/.

340 Hevesy, G. *Adventures in Radioisotope Research.* (Pergamon Press, 1962).

341 Allen, S. R. *Niels Bohr: The Man, His Science, and the World They Changed.* (Alfred A. Knopf, 1966).

342 Dawidowicz, L. S. *The War against the Jews 1933–1945.* (Holt, Rinehart & Winston, 1975).

343 Tenenbaum, J. Auschwitz in retrospect: the self-portrait of Rudolf Hoess, Commander of Auschwitz. *Jewish Social Studies* **15**, 203–36 (1953).

344 Witschi, H. Some notes on the history of Haber's Law. *Toxicological Sciences* **50**, 164–68 (1999).

345 *Law Reports of Trials of War Criminals: Case No. 9 The Zyklon B Case, Trial of Bruno Tesch and Two Others, British Military Court, Hamburg, 1st–8th March, 1946.* (United Nations War Crimes Commission, 1947).

346 *Convention (IV) respecting the Laws and Customs of War on Land and Its Annex: Regulations concerning the Laws and Customs of War on Land. The Hague, 18 October 1907.* (1907).

347 Lutz F. Haber (1921–2004). Division of History of Chemistry of the American Chemical Society (2006).

348 Haber, L. F. *The Poisonous Cloud: Chemical Warfare in the First World War.* (Oxford University Press, 1986).

349 Simmons, J. S. How magic is DDT? *Saturday Evening Post* **217** (1945).

350 *Pearl Harbor: America's Call to Arms.* (Life Books, 2011).

351 Leuchtenburg, W. E. *The Life History of the United States, Vol. 11: 1933–1945: New Deal and War.* (Time-Life Books, 1964).

352 *Reports of General MacArthur: The Campaigns of MacArthur in the Pacific, Vol. 1.* (General Headquarters, US Army Forces, Far East, 1966).

353 Leckie, R. *Strong Men Armed: The United States Marines against Japan.* (Da Capo Press, 1962).

354 Bray, R. S. *Armies of Pestilence: The Impact of Disease on History.* (Barnes & Noble, 1996).

355 *Encyclopedia of Pestilence, Pandemics, and Plagues.* (Greenwood Press, 2008).

356 Greenwood, J. T. The fight against malaria in the Papua and New Guinea campaigns. *Army History* **59**, 16–28 (2003).

357 Joy, R. J. T. Malaria in American troops in the South and Southwest Pacific in World War II. *Medical History* **43**, 192–207 (1999).

358 Griffin, A. R. *Out of Carnage.* (Howell, Soskin, Publishers, 1945).

359 Laurence, W. L. New drugs to combat malaria are tested in prisons for Army. *New York Times* **March 5** (1945).

360 Geissler, E., & Guillemin, J. German flooding of the Pontine Marshes in World War II: biological warfare or total war tactic? *Politics and the Life Sciences* **29**, 2–23 (2010).

361 McCormick, A. O. Undoing the German campaign of the mosquito. *New York Times* **September 13** (1944).

362 Jacobsen, A. *Operation Paperclip.* (Little, Brown & Co., 2014).

363 Perkins, J. H. Reshaping technology in wartime: the effect of military goals on entomological research and insect-control practices. *Technology and Culture* **19**, 169–86 (1978).

364 Kaempffert, W. DDT, the Army's insect powder, strikes a blow against typhus and for pest control. *New York Times* **June 4** (1944).

365 Zeidler, O. Verbindungen von Chloral mit Brom- und Chlorbenzol. *Berichte der deutschen chemischen Gesellschaft* **7**, 1180–81 (1874).

366 *Nobel Lectures, Physiology or Medicine, 1942–1962.* (Elsevier Publishing Co., 1964).

367 Müller, P. H. Dichloro-diphenyl-trichloroethane and newer insecticides. Nobel Lecture **December 11** (1948).

368 Knipling, E. F. DDT insecticides developed for use by the armed forces. *Journal of Economic Entomology* **38**, 205–7 (1945).

369 Annand, P. N. Tests conducted by the Bureau of Entomology and Plant Quarantine to appraise the usefulness of DDT as an insecticide. *Journal of Economic Entomology* **37**, 125–26 (1944).

370 Gardner, L. R. Fifty years of development in agricultural pesticidal chemicals. *Industrial and Engineering Chemistry* **50**, 48–51 (1958).

371 Gahan, J. B., Travis, B. V., & Lindquist, A. W. DDT as a residual-type spray to control disease-carrying mosquitoes: laboratory tests. *Journal of Economic Entomology* **38**, 236–40 (1945).

372 DDT for peace. *New York Times* **July 15** (1945).

373 Bishopp, F. C. Present position of DDT in the control of insects of medical importance. *American Journal of Public Health and the Nation's Health* **36**, 593–606 (1946).

374 Kaempffert, W. The year saw many discoveries and advances hastened by the demands of the war. *New York Times* **December 31** (1944).

375 Typhus in Naples checked by Allies. *New York Times* **February 22** (1944).

376 The conquest of typhus. *New York Times* **June 4** (1944).

377 Simmons, J. S. Preventive medicine in the Army. In *Doctors at War* (ed. M. Fishbein). (E. P. Dutton & Co., Inc., 1945).

378 Typhus blockade is set up at Rhine. *New York Times* **April 10** (1945).

379 Long, T. Child evacuation stirs Berlin fear. *New York Times* **October 29** (1945).

380 Gahan, J. B., Travis, B. V., Morton, P. A., & Lindquist, A. W. DDT as a residual-type treatment to control *Anopheles quadrimaculatus*: practical tests. *Journal of Economic Entomology* **38**, 231–35 (1945).

381 Soper, F. L., Knipe, F. W., Casini, G., Riehl, L. A., & Rubino, A. Reduction of *Anopheles* density effected by the preseason spraying of building interiors with DDT in kerosene, at Castel Volturno, Italy, in 1944–45 and in the Tiber Delta in 1945. *American Journal of Tropical Medicine and Hygiene* **27**, 177–200 (1947).

382 Army to use DDT powder on malaria mosquitos. *New York Times* **August 1** (1944).

383 Kirk, N. T. School of battle for doctors. *New York Times* **November 26** (1944).

384 Montagu, M. F. A. Calling all doctors. *New York Times* **May 20** (1945).

385 Text of the review by Prime Minister Churchill on military and political situations, speech in the House of Commons, September 28, 1944. *New York Times* **September 29** (1944).

386 Spraying an island. *New York Times* **December 24** (1944).

387 Saipan cleansed. Airplanes spraying island with DDT, killing every insect. *New York Times* **December 3** (1944).

388 Shalett, S. Plane's-eye view of the Pacific War. *New York Times* **January 14** (1945).

389 Container outlook for 1945 improved. *New York Times* **December 6** (1944).

390 More woolens set for civilian use. *New York Times* **June 30** (1945).

391 DDT cost cut 40% since July. *New York Times* **December 29** (1944).

392 Russell, P. F. Lessons in malariology from World War II. *American Journal of Tropical Medicine and Hygiene* **26**, 5–13 (1946).

393 Pacific bugs face rain of DDT bombs. *New York Times* **August 18** (1945).

394 Fishbein, M. *Doctors at War*. (E. P. Dutton & Co., Inc., 1945).

395 Planes to fight malaria. *New York Times* **August 9** (1945).

396 Chemists say DDT could save 1 to 3 million lives each year. *New York Times* **August 29** (1945).

397 Insect-killing fog is tested at beach. *New York Times* **July 9** (1945).

398 Long Island beaches rid of insects by DDT. *New York Times* **July 25** (1945).

399 Public to receive DDT insecticide. *New York Times* **July 27** (1945).

400 Russell, E. P. "Speaking of annihilation": mobilizing for war against human and insect enemies, 1914–1945. *Journal of American History* **82**, 1505–29 (1996).

401 DDT mixed in wall paint keeps flies from rooms. *New York Times* **August 23** (1945).

402 DDT repels barnacles. *New York Times* **July 17** (1945).

403 Notes of science: DDT spray. *New York Times* **July 22** (1945).

404 Use of big guns urged to kill Jersey "skeeters." *New York Times* **March 31** (1945).

405 Flies on Mackinac Island extinguished with DDT. *New York Times* **August 10** (1945).

406 Spray DDT in polio area. *New York Times* **August 14** (1945).

407 DDT sprayed over Rockford, Ill., in test of power to halt polio. *New York Times* **August 20** (1945).

408 Use of DDT plane sought by Jersey. *New York Times* **August 21** (1945).

409 Macy's display advertisement. *New York Times* **August 19** (1945).

410 Bloomingdale's Sky Greenhouse display advertisement. *New York Times* **September 9** (1945).

411 This time it is the elephant that gets a spraying. *New York Times* **September 13** (1945).

412 Bomb-type insecticide dispensers slated to be more available in the stores here. *New York Times* **September 18** (1945).

413 U.S. tells how to use DDT against insects; plans drive on fraudulent mixtures. *New York Times* **September 25** (1945).

414 Advertising news and notes. *New York Times* **October 15** (1945).

415 Spollen, P. Choosing gifts for gardeners. *New York Times* **November 25** (1945).

416 Macy's display advertisement. *New York Times* **October 1** (1945).

417 Mayor gets fund for Morris talk. *New York Times* **October 14** (1945).

418 Tojo's jail. *New York Times* **October 14** (1945).

419 Lyle, C. Achievements and possibilities in pest eradication. *Journal of Economic Entomology* **40**, 1–8 (1947).

420 Rudd, R. *Pesticides and the Living Landscape*. (University of Wisconsin Press, 1964).

421 Davis, F. R. *Banned: A History of Pesticides and the Science of Toxicology.* (Yale University Press, 2014).

422 Garnham, C., Heisch, R. B., Harper, J. O., & Bartlett, D. DDT versus malaria: a successful experiment in malaria control by the Kenya Medical Department. Film. East African Sound Studios (1947).

423 Macchiavello, A. Plague control with DDT and "1080": results achieved in a plague epidemic at Tumbes, Peru, 1945. *American Journal of Public Health* **36**, 842–54 (1946).

424 Lal, H. Of men and mosquitoes. *Scientist and Citizen* **8**, 1–5 (1965).

425 MacGillivray, A. *Words That Changed the World: Rachel Carson's Silent Spring.* (Ivy Press, Ltd., 2004).

426 *Trials of War Criminals before the Nuernberg Military Tribunals under Control Council Law No. 10, Vol. 7: Nuernberg, October 1946–April 1949.* (US Government Printing Office, 1953).

427 *Arms and the Men.* (Doubleday, Doran & Co., Inc., 1934).

428 Engelbrecht, H. C., & Hanighen, F. C. *Merchants of Death.* (Dodd, Mead & Co., 1934).

429 Barnard, E. Academic freedom demanded by NEA. *New York Times* **July 5**, 5 (1935).

430 Roosevelt moves to gain control of arms traffic. *Spokane Daily Chronicle* **May 18**, 1 (1934).

431 Harris, R., & Paxman, J. *A Higher Form of Killing: The Secret History of Chemical and Biological Warfare.* (Random House Trade Paperbacks, 2007).

432 Military Tribunal VI, Judgment of the Tribunal, Trial 6—I. G. Farben Case. (1948).

433 Bacon, R. *De secretis operibus artis et naturae et de nullitate magiae.* (Frobeniano, 1618).

434 Timperley, C. M. *Best Synthetic Methods Organophosphorus (V) Chemistry.* (Academic Press, 2015).

435 Schrader, G. *The Development of New Insecticides and Chemical Warfare Agents.* British Intelligence Objectives Sub-Committee (B.I.O.S.) Final Report No. 714, Item No. 8, Presented by S. A. Mumford and E.A. Perren, Black List Item 8, Chemical Warfare (BIOS Trip No. 1103). (1945).

436 Baader, G., Lederer, S. E., Low, M., Schmaltz, F., & von Schwerin, A. Pathways to human experimentation, 1933–1945: Germany, Japan, and the United States. *Osiris,* **2d series, 20** (*Politics and Science in Wartime: Comparative International Perspectives on the Kaiser Wilhelm Institute*), 205–31 (2005).

437 Preuss, J. The reconstruction of production and storage sites for chemical warfare agents and weapons from both world wars in the context of assessing former munitions sites. In *One Hundred Years of Chemical Warfare: Research, Deployment, Consequences* (ed. B. Friedrich et al.). (SpringerOpen, 2017).

438 Perry, M., & Schweitzer, F. M. *Antisemitism: Myth and Hate from Antiquity to the Present*. (Palgrave Macmillan, 2002).

439 Better pest control. *Science News Letter* **60**, 340 (1951).

440 Sidell, F. R. Nerve agents. In *Medical Aspects of Chemical and Biological Warfare* (ed. F. R. Sidell, E. T. Takafuji, & D. R. Franz). (Office of the Surgeon General, Department of the Army, United States of America, 1997).

441 Schrader, G. The development of new insecticides. British Intelligence Objectives Sub-Committee (B.I.O.S.), B.I.O.S. Trip No. 1103: B.I.O.S. Target Nos. 08/85, 8/12, 08/159, 8/59(B). (1945).

442 Robinson, J. P. *The Problem of Chemical and Biological Warfare, Vol. 1: The Rise of CB Weapons*. (Stockholm International Peace Research Institute, 1971).

443 Schmaltz, F. Chemical weapons research on soldiers and concentration camp inmates in Nazi Germany. In *One Hundred Years of Chemical Warfare: Research, Deployment, Consequences* (ed. B. Friedrich et al.). (SpringerOpen, 2017).

444 Corey, R. A., Dorman, S. C., Hall, W. E., Glover, L. C., & Whetstone, R. R. Diethyl 2-chlorovinyl phosphate and dimethyl 1-carbomethoxy-1-propen-2-yl phosphate—two new systemic phosphorus pesticides. *Science* **118**, 28–29 (1953).

445 Metcalf, R. L. The impact of the development of organophosphorus insecticides upon basic and applied science. *Bulletin of the Entomological Society of America* **5.1**, 3–15 (1959).

446 Shaw, G. B. *Man and Superman: A Comedy and a Philosophy*. (Archibald Constable & Co., Ltd., 1903).

447 Carson, R. *Silent Spring*. (Houghton Mifflin Co., 1962).

448 Howard, L. O. The war against insects. *Chemical Age* **30**, 5–6 (1922).

449 Russell, L. M. Leland Ossian Howard: a historical review. *Annual Review of Entomology* **23**, 1–15 (1978).

450 Howard, L. O. *Mosquitoes, How They Live; How They Carry Disease; How They Are Classified; How They May Be Destroyed*. (McClure, 1901).

451 Andrews, J. M., & Simmons, S. W. Developments in the use of the newer organic insecticides of public health importance. *American Journal of Public Health* **38**, 613–31 (1948).

452 De Bach, P. The necessity for an ecological approach to pest control on citrus in California. *Journal of Economic Entomology* **44**, 443–47 (1951).

453 Melander, A. L. Can insects become resistant to sprays? *Journal of Economic Entomology* **7**, 167–72 (1914).

454 Quayle, H. J. The development of resistance to hydrocyanic acid in certain scale insects. *Hilgardia* **11**, 183–210 (1938).

455 Quayle, H. J. Are scales becoming resistant to fumigation? *California University Journal of Agriculture* **3**, 333–34, 358 (1916).

456 Livadas, G. A., & Georgopoulos, G. Development of resistance to DDT by *Anopheles sacharovi* in Greece. *Bulletin of the World Health Organization* **8**, 497–511 (1953).

457 *Conference on Insecticide Resistance and Insect Physiology.* (Division of Medical Sciences, National Research Council at the request of the Army Medical Research and Development Board, 1952).

458 Clement, R. C. The pesticides controversy. *Boston College Environmental Affairs Law Review* **2**, 445–68 (1972).

459 Curran, C. H. DDT: the atomic bomb of the insect world. *Natural History* **54**, 401–5, 432 (1945).

460 Teale, E. W. DDT: it can be a boon or a menace. *Nature Magazine* **38**, 120 (1945).

461 Cottam, C., & Higgins, E. *DDT: Its Effects on Fish and Wildlife. US Department of the Interior, Fish and Wildlife Service, Circular* **11** (1946).

462 Barker, R. J. Notes on some ecological effects of DDT sprayed on elms. *Journal of Wildlife Management* **22**, 269–74 (1958).

463 Graham, F. J. *Since Silent Spring.* (Houghton Mifflin Co., 1970).

464 Strother, R. S. Backfire in the war against insects. *Reader's Digest* **74**, 64–69 (1959).

465 Pesticides are good friends, but can be dangerous enemies if used by zealots. *Saturday Evening Post* **September 2**, 8 (1961).

466 Murphy, P. C. *What a Book Can Do: The Publication and Reception of Silent Spring.* (University of Massachusetts Press, 2005).

467 Lytle, M. H. *The Gentle Subversive.* (Oxford University Press, 2007).

468 Tennyson, A. *Poems.* (W. D. Ticknor, 1842).

469 Carson, R. Undersea. *Atlantic Monthly* **160**, 322–25 (1937).

470 Carson, R. *Under the Sea-Wind: A Naturalist's Picture of Ocean Life.* (Simon & Schuster, 1941).

471 Brooks, P. *The House of Life: Rachel Carson at Work.* (Houghton Mifflin Co., 1972).

472 Carson, R. *The Sea around Us.* (Oxford University Press, 1951).

473 Leonard, J. N. And his wonders in the deep. *New York Times* **July 1** (1951).

474 Quaratiello, A. R. *Rachel Carson: A Biography.* (Greenwood Press, 2004).

475 Carson, R. *The Edge of the Sea.* (Houghton Mifflin, 1955).

476 Poore, C. Books of the Times. *New York Times* **October 26**, 29 (1955).

477 Galbraith, J. K. *The Affluent Society.* (Hamish Hamilton, 1958).

478 Japanese bid U.S. curb atom tests. *New York Times* **April 1**, 26 (1954).

479 Cow contamination by fall-out studied. *New York Times* **April 14**, 6 (1959).

480 Reiss, L. Z. Strontium-90 absorption by deciduous teeth. *Science* **134**, 1669–73 (1961).

481 New group to seek "SANE" atom policy. *New York Times* **November 15**, 54 (1957).

482 Hunter, M. Arms race opposed—response cheers head of "strike." *New York Times* **November 22**, 4 (1961).

483 Finney, J. W. U.S. atomic edge believed in peril. *New York Times* **October 27**, 1, 7 (1962).

484 Dean, C. Cranberry sales curbed; U.S. widens taint check. *New York Times* **November 11**, 1, 29 (1959).

485 Text of Eisenhower's farewell address. *New York Times* **January 18**, 22 (1961).

486 Mintz, M. "Heroine" of FDA keeps bad drug off market. *Washington Post* **July 15**, 1 (1962).

487 Diamond, E. The myth of the "Pesticide Menace." *Saturday Evening Post* **28 September**, 16, 18 (1963).

488 Decker, G. C. Pros and cons of pests, pest control and pesticides. *World Review of Pest Control* **1**, 6–18 (1962).

489 Keats, J. *The Poetical Works of John Keats*. (DeWolfe, Fiske & Co., 1884).

490 President's Science Advisory Committee. *Use of Pesticides*. (US Government Printing Office, 1963).

491 Lee, J. M. "Silent Spring" is now noisy summer. *New York Times* **July 22**, 87, 97 (1962).

492 Rachel Carson's warning. *New York Times* **July 2**, 28 (1962).

493 Lear, L. *Rachel Carson: Witness for Nature*. (Henry Holt & Co., 1997).

494 Lutts, R. H. Chemical fallout: *Silent Spring*, radiactive fallout, and the environmental movement. In *And No Birds Sing: Rhetorical Analyses of Rachel Carson's* Silent Spring (ed. C. Waddell). (Southern Illinois University Press, 2000).

495 The desolate year. *Monsanto Magazine* **October**, 4–9 (1962).

496 Bean, W. B. The noise of *Silent Spring*. *Archives of Internal Medicine* **112**, 308–11 (1963).

497 Darby, W. J. Silence, Miss Carson. *Chemical and Engineering News* **October 1**, 60–63 (1962).

498 Stare, F. J. Some comments on *Silent Spring*. *Nutrition Reviews* **21**, 1–4 (1963).

499 Wyant, W. K. J. Bug and weed killers: blessings or blights? *St. Louis Post Dispatch* **July 28**, 2–3 (1962).

500 White-Stevens, R. H. Communications create understanding. *Agricultural Chemicals* **17**, 34 (1962).

501 Biology: pesticides: the price for progress. *Time* **September 28**, 45–47 (1962).

502 Hayes, W. J. J., Durharm, W. F., & Cueto, C. J. The effect of known repeated oral doses of chlorophenothane (DDT) in man. *Journal of the American Medical Association* **162**, 890–97 (1956).

503 Vogt, W. On man the destroyer. *Natural History* **72**, 3–5 (1963).

504 Hawkins, T. R. Re-reading *Silent Spring*. *Environmental Health Perspectives* **102**, 536–37 (1994).

505 Leonard, J. N. Rachel Carson dies of cancer; "Silent Spring" author was 56. *New York Times* **April 15** (1964).

506 Critic of pesticides: Rachel Louise Carson. *New York Times* **June 5**, 59 (1963).

507 Kraft, V. The life-giving spray. *Sports Illustrated* **November 18**, 22–25 (1963).

508 Agassiz, L. Professor Agassiz on the *Origin of Species*. *American Journal of Science* **30**, 143–47, 149–50 (1860).

509 Paine, T. *Rights of Man: Answer to Mr. Burke's Attack on the French Revolution.* (J. S. Jordan, 1791).

510 Atkinson, B. Rachel Carson's "Silent Spring" is called "The Rights of Man" of our time. *New York Times* **April 2**, 44 (1963).

511 *CBS Reports: The Silent Spring of Rachel Carson*. TV program. (1963).

512 Vollaro, D. R. Lincoln, Stowe, and the "little woman/great war" story: the making, and breaking, of a great American anecdote. *Journal of the Abraham Lincoln Association* **30**, 18–34 (2009).

513 Carson, R. Rachel Carson answers her critics. *Audubon Magazine* **September**, 262–65 (1963).

514 Free, A. C. *Animals, Nature & Albert Schweitzer*. (Flying Fox Press, 1982).

515 White, E. B. Notes and comment. *New Yorker* **May 2** (1964).

516 Lear, L. Introduction. In *Silent Spring* (R. Carson). (Houghton Mifflin Company, 2002).

517 Kinkela, D. *DDT and the American Century*. (University of North Carolina Press, 2011).

518 Cecil, P. F. *Herbicidal Warfare. The Ranch Hand Project in Vietnam*. (Praeger Scientific, 1986).

519 *Veterans and Agent Orange: Health Effects of Herbicides Used in Vietnam*. (National Academy Press, 1994).

520 Meselson, M. From Charles and Francis Darwin to Richard Nixon: the origin and termination of anti-plant chemical warfare in Vietnam. In *One Hundred Years of Chemical Warfare: Research, Deployment, Consequences* (ed. B. Friedrich et al.). (Springer Open, 2017).

521 Lewis, J. G. On Smokey Bear in Vietnam. *Environmental History* **11**, 598–603 (2006).

522 Primack, J., & von Hippel, F. *Advice and Dissent: Scientists in the Political Arena*. (Basic Books, Inc., 1974).

523 Meselson, M. S., Westing, A. H., Constable, J. D., & Cook, J. E. Preliminary report of the Herbicide Assessment Commission of the American Association for the Advancement of Science. **December 30** (1970).

524 Westing, A. H. Herbicides as agents of chemical warfare: their impact in relation to the Geneva Protocol of 1925. *Boston College Environmental Affairs Law Review* **1**, 578–86 (1971).

525 Konrad, K. Lois Gibbs: grassroots organizer and environmental health advocate. *American Journal of Public Health* **101**, 1558–59 (2011).

526 Gibbs, L. *Love Canal: My Story.* (State University of New York Press, 1983).

527 Brown, M. *Laying Waste: The Poisoning of America by Toxic Chemicals.* (Pantheon Books, 1979).

528 Chávez, C. What is the worth of a man or a woman? Speech at Pacific Lutheran University, Tacoma, Washington. (1989).

529 Goulson, D. An overview of the environmental risks posed by neonicotinoid insecticides. *Journal of Applied Ecology* **50**, 977–87 (2013).

530 von Hippel, A. R. *Life in Times of Turbulent Transitions.* (Stone Age Press, 1988).

531 Reisman, A. Einstein the savior. *Jewish Magazine* **June** (2010).

532 Reisman, A. Jewish refugees from Nazism, Albert Einstein, and the modernization of higher education in Turkey (1933–1945). *Aleph* **7**, 253–81 (2007).

533 Reisman, A. What a freshly discovered Einstein letter says about Turkey today. *History News Network* **November 20** (2006).

534 Teller, E., & Shoolery, J. *Memoirs. A Twentieth-Century Journey in Science and Politics.* (Perseus Publishing, 2001).

535 von Hippel, F. N. Arthur von Hippel: the scientist and the man. *MRS Bulletin* **30**, 838–44 (2005).

536 Smith, A. K. *A Peril and a Hope. The Scientists' Movement in America 1945–47.* (MIT Press, 1965).

537 *The Manhattan Project: A Documentary Introduction to the Atomic Age.* (Temple University Press, 1991).

Index

Page numbers in italics indicate figures.

American Public Health Association, 79
American Revolution, 34
aminotriazole, 263
amiton, 243
ammonia, 129, 137, 141–42, 149, 153, 158, 164–65, 230, 299
aniline, 151, 187
Animal Welfare Institute, 282
Anopheles, 31, 42–43, 47–49, *52*, 198, 202, 287–88
Anschluss, 317
anthrax, 17, 35
antifreeze, 143
antigas ointment, 233, 235
antisemitism, 140, 142, 171, 299–301, 304
ants, 161, 185, 219, 255–56, 261, 315
Aoyama, Tenemichi, 111, 113, 117
aphids, 21–22, 226, 245, 317
Apollo Smintheus, 98
Arbeit macht frei ("Work sets you free") motto, 221
Archelaus, 14
Archives of Internal Medicine (journal), 277–78
Aristotle, 14
Arlington National Cemetery, 79
Army Weapons Office, 227
Arrhenius, Svante, 165
arsenate of lead, 25–26
arsenic, 12, 25–26, 128, 153–56
Aryanism, 171–72, 302
Asibi strain, 83–84, 86
asphyxiation, 137, 144, 168
assassins, 59, 79, 134, 170, 172, 233, 312
atabrine, 188–90, 199–200
Athens, 133–34
Atlantic Monthly (magazine), 258
atomic hydrogen, 139
Auschwitz, 97, 179–83, 220–24

Austria, 94, 96, 191, 218
automobiles, ix–xi, 139, 143, 170, 226, 300
avian malaria, 44–47
Avignon, 105–6

bacillus: Black Death and, 112–13, 115, 117–18, 120, 122, 124–25; Kitasato and, 112–13, 115, 117–19; tetanus, 112; yellow fever and, 68, 74, 78–79; Yersin and, 118
Bacillus icteroides, 68, 78
Bacon, Francis, 89–90
Bacon, Roger, 225
Bacot, Arthur, 124–25, 127, 129
bacteria, 31, 35, 78, 87, 112, 116–17, 123, 225, 239
balance of nature, 251, 253, 259, 267, 275, 280, 288, 296, 318
bald eagle, 255
BASF, 216
Bastianelli, Giuseppe, 41
Bataan, 185
Bauhin, Gaspard, 4
Bayer, 216, *227*, 246
Beauperthuy, Louis-Daniel, 64, 86
bedbugs, 166, 194, 253
bees, 14, 283, 296
beetles, 25, 51, 194, 245–46, 256
Belgium, 7, 144, 219
Bellevue Hospital Medical College, 67
benzene, 129
Berkeley, M. J., 12–15, 18–19
Better Business Bureau, 208
"Better Living through Chemistry" (DuPont), 272
BHC, 212
Bhopal disaster, 295
Bickenbach, Otto, 239
Bignami, Amico, 41, 47
Bikini atoll, 262

bloodletting, 61–62, 106–7
bluebottle flies, 192–93
Bohr, Niels, 147, 174–76, 299, 307–9, 312
Bolivia, 288
boll weevil, 153
Bolshevik Red Army, 97
Book-of-the-Month Club, 260, 269
Booth, John Wilkes, 37
Bordeaux Mixture, 22–26
Bormann, Martin, 232
Born, Max, 302, 304–5, 315
Bosch, Carl, 217–19, 304
Boston, 9, 59, 63
Boston Globe (newspaper), 260
Botrytis infestans, 18
Bouhler, Philipp, 178
Brack, Victor, 178
Brandt, Karl, 178
Brauner, 305
Bréant Prize, 36
breast milk, 272, 287
Brill, Nathan, 97–98
Brill-Zinsser disease, 98
British Admiralty, 134
British Imperial Army, 199
British Indian Commission, 123
British Intelligence, 175
British Intelligence Objectives Sub-Committee (BIOS), 237–38, 242
British Medical Journal, 110–11, 115–17
British Military Court, 181
British Queen, The (potato), 11
Brockovich, Erin, 294–95
bromide, 143, 168–69
bromine, ix
Brooks, Paul, 268, 278–79, 284
bubonic plague. *See* Black Death
bug bombs, 201–2, 208, 276
Building 144, 231

Bulletin of the Atomic Scientists, 312
bumblebees, 296
Bunsen Society, 141
Burma, 189, 199
Burnside, 137
buzzards, 44, 161

cadavers, 60, 101, 114–15
Caffa, 101–2
calcium arsenate, 153
Calliphora vomitoria, 192–93
Cambodia, 112, 288
camp fever. *See* typhus
camphor, 128–29
Camp Lazear, 72–73
Camp Overcast, 240
Canada, 8, 58, 145, 183, 197
cancer, 225, 262–63, 266, 279, 290
Cantlie, James, 108, 119–21
Caracalla, 32
carbaryl, 295
carbolic acid, 63, 128–29
carbon monoxide, 178–80
Carey, Matthew, 54–55
Carmelites, 66
Carroll, James, 64, 67, 69–73, 77–78, 80, 85, 87
Carson, Rachel, xi–xii, 249; Arthur von Hippel and, 311; awards of, 260, 282–83; background of, 257–58, 261–62, 265–66; chemical industry and, 286–87; continued chemical hazards after death of, 287–98; critics of, 269–79, 286–87; DDT and, 259, 267, 269, 272, 275–76, 282, 287, 297; death of, 283–84, 287, 317; doctors and, 266, 279; *The Edge of the Sea* and, 261, 283; fact vetting by, 268; good reviews of, 259–60; Government Operations Subcommittee and, 280–81; Gug-

chemicals of war (*cont.*)
308; organophosphates and, 212,
226, 231, 234, 236–46, 254, 287,
296; pesticides and, 147, 153–54,
163, 168, 198, 200, 205, 211–14,
226, 242–46; physicians and, 178,
202, 222, 237; prisoners and, 179–
80, 188, 190–91, 197, 220–23,
229, 235; profits and, 160, 163,
167–69, 214–16, 221, 226, 297;
rats and, 145, 147, 156, 161–62,
196, 225, 244; Robbers' Lair and,
236–37; Rocky Mountain Arsenal
and, 205, 293–94; safety and, 154,
168, 174, 176, 180, 182, 188, 195–
96, 212, 216, 242–43; synthesis
and, 130, 137–38, 141–42, 150,
154–56, 165, 187–96, 217, 219,
226, 230–31, 235, 238, 242, 246,
299; toxicity and, 133, 150, 153,
163, 192–94, 226–27, 230–31,
237–38, 243–45; Treaty of Versailles and, 168–69; typhus and,
154; US Chemical Warfare Service
and, 147, 154, 156, 160–63, 166,
199, 205, 212, 225, 237, 240–41,
289; weapons of mass destruction
and, 144, 293; Wöhler and, 137–
38, 299; women and, 157, 179,
222, 234; World War I and, 130,
133, 138, 142, 147–49, 156, 159–
60, 166–70, 177–84, 188, 197,
199–200, 215–16, 235; World
War II and, 181, 188, 216, 219,
235, 246. *See also specific chemicals*
Chemie Grünenthal, 264–65
chiggers, 185
chikungunya, 85
childbed fever, 276
children: birth defects and, 265, 290–
92; Black Death and, 103, 112;

Bohr and, 174–75; breast milk
and, 272, 287; Carson and, 258;
chemicals of war and, 157; DDT
and, 257; endrin and, 287; evacuation of, 197–98; experiments on,
190; herbicides and, 292; Höss
and, 182; I. G. Farben and, 222;
infant mortality and, 174, 276,
307–8, 318; leaded gasoline and,
x; malaria and, 48; Nazis and,
178–79, 182–83, 190, 222, 317;
Nernst and, 139, 178; orphaned,
55; Pasteur and, 17; stillbirths
and, 287; strontium-90 and, 262;
thalidomide and, 264–65, 291;
typhus and, 95–96; White and,
261; yellow fever and, 55, 65–68,
80; Zyklon and, 183
Chile, 141, 153
chimpanzees, 83, 93, 239
China, 5, 31, 37, 108–12, 118, 120,
126
chlordane, 212, 253, 256, 269
chlorine, ix, 137, 144–47, 149, 162,
168–69, 246, 271, 295
chloroethyl alcohol, 226
chlorofluorocarbon (CFC), x
chloroform, 154, *155*
chloropicrin, 154
cholera, 11, 17, 21, 35, 41, 68, 88
Christians, 101–4, 140, 170, 173, 217
Christmas, 143, 208
Christmas Island, 48
chromium-6, 295
Churchill, Winston, 136, 159, 199,
234–35, 258, 279, 284, 305
cinch bug, 156
cinchona trees, 34–35, 61, 118, 128,
186–87
cinchonine, 35, 128
Civil Service Law, 171

Clarkson, Matthew, 57
cleaning products, 297
Clement VII, Pope, 90, 104, 106
clothing, 307; Black Death and, 106,
    121; chemicals of war and, 167,
    169, 182, 208; Irish poverty and,
    9; typhus and, 91–93, 95; yellow
    fever and, 59, 63, 72–74, 79
coal tar, 129, 153, 187–88
Cochrane, Thomas, 134–36
Cold War, 242, 271
Colombia, 82, 287–88
colonies, 4, 163–64; Africa and, 33,
    35, 37, 48, 50, 53, 86, 212; Black
    Death and, 113; malaria and,
    112; postwar expansion of, 159;
    smallpox and, 112; yellow fever
    and, 58, 60
colony collapse disorder, 296
Colorado potato beetles, 25, 245–46
comas, 89, 111, 195, 252
Combined Intelligence Objectives Sub-
    committee (CIOS), 240–41
compound 2,4-D, 289–90
compound 2,4,5-T, 289–92
Comprehensive Environmental
    Response, Compensation, and Li-
    ability Act of 1980, 294
Compton, Arthur Holly, 311–13
Compton, Karl, 309–11
concentration camps: Auschwitz,
    97, 179–83, 220–24; Birkenau,
    180–81, 221–23; carbon monoxide
    and, 178–80; chemicals of war
    and, 178–79, 181, 190, 220–22,
    229, 235–36, 239; Dachau, 190–
    91; Eichmann and, 179–81; food
    and, 222–23; gas chambers and,
    167, 169, 178–81, 194, 221–23,
    317; Höss and, 179–80, 182; I. G.
    Farben and, 220–24, 229, 235–36,

239; incinerators of, 180; Nazis
    and, 97, 129, 178–79, 181, 190,
    220–22, *229*, 235–36, 239, 314;
    Zyklon and, 178–81, 194, 220
concrete, 293
Confederates, 59, 136–37
Congressional Record, 216
Constable, John, 292
Constantinople, 101
consumerism, 208, 262
convulsions, 56, 152, 252, 282
Cook, Robert, 292
copper, 12, 22–25, 51, 134
corn borers, 256
Corsica, 101–2, 201
cotton, 152–53
Counter Intelligence Corps, 241
Courant, Richard, 302
cranberries, 263
Creation, 20
Crimea, 91, 101–2, 135
Crossbow, 232
crows, 44, 161
Crusades, 134
cryolite, 211
Cuba, 51, 63–67, 69, 71, 72, 75–82, 95,
    250, 262
culicicides, 49, 51, 81
cyanic acid, 137, 299
cyanide, 136, 154, 182, 226–28, 232
cyclohexanone, 202
Cyklon, 168
Czechoslovakia, 218

Dachau, 190–91
Daladier, Edouard, 218
Dance of Death, 107–8
Danish resistance, 175
Dark Ages, 100–101, 280
Darwin, Charles, 12, 15, 20, 23, 253,
    278–79

Carson and, 266, 279; famine and, 9–10; malaria and, 186, 190–91; Nazis and, 190–91, 240; Operation Paperclip and, 191, 240; potato blight and, 9–10; thalidomide and, 264–65; typhus and, 9, 92, 96; war crimes and, 190–91; yellow fever and, 54, 60–61, 65, 67, 69

dogs, 17, 65, 68, 83, 149, 202, 227, 239

Douglas, William O., 267, 269

Dow Chemical Company, 241, 257, 272, 291

dry rot, 7, 25

DuBridge, Lee, 290–91

Dugway Proving Ground, 243–44, 293

DuPont, ix–x, 151, 201, 216, 256, 272

Dustbin, 237, 241, 242

Dutch East Indies, 184–85, 191

dyes, 138, 150–51, 164, 187–88, 216, 218

dysentery, 33, 88, 91, 112, 185

Earl of Clare, 7

early blight, 25

ecological issues, xi–xii; balance of nature and, 251, 253, 259, 267, 275, 280, 288, 296, 318; Carson and, 257 (*see also* Carson, Rachel); conquest of nature and, 249–50; damage to wildlife and, 251, 254–55, 263, 275, 278, 282, 296; defoliants and, 288–91; famine and, 249, 270–72, 276, 281, 286, 297, 299; fleas and, 253, 263; flies and, 251, 253–54, 256, 276; food and, 249, 259, 263–65, 269–71, 274, 276, 280, 283, 286–92; Geneva Protocol and, 293; Germany and, 264–65; insecticides and, 250–55, 268–69, 274–76, 280, 287,

295–96; mice and, 256, 271, 290; military-industrial complex and, 264; mortality and, 255, 276, 282; mosquitoes and, 250–53, 255, 271, 287–88, 292; nuclear power and, 262, 271, 278, 289, 296; ozone and, x, 202, 296; pesticides and, 251–52, 255–57, 261–65, 268–73, 277–84, 287–90, 295–97, 317; physicians and, 265; radiation and, x, 262, 266, 271; rats and, 263, 271; resistance and, 253–56, 281, 288, 296–97; safety and, 254, 264, 269, 295; *Silent Spring* and, 262, 264, 267–84, 287–90, 296–97; Superfund and, 294; synthetics and, 264, 287, 289, 291, 295–96; toxicity and, 251–52, 255, 281–82, 287, 290, 294–97; US Fish and Wildlife Service and, 254–55, 259; US Food and Drug Administration and, 263–65, 280, 291; women and, 269, 277–78, 283, 287; World War I and, 249, 271; World War II and, 250–51, 282, 286, 289, 293

Ecology Center in Berkeley, 287

Economist (journal), 276

Edge of the Sea, The (Carson), 261, 283

Edison, Thomas, 138–39

Egypt, 51, 118, 127, 129

Eichmann, Adolf, 179–81

Einstein, Albert: Born and, 305; calls for assassination of, 170; Franck and, 147, 301, 314–16; Inonu letter of, 306; Nobel Prize of, 143, 301; resignation of, 171, 178; as traitor of Nazism, 174

Eisenhower, Dwight D., 224, 235, 263–64, 284–85

electricity, 11–12, 138, 165, 208, 221, 226, 294, 301, 308–11

253; ecological issues and, 253, 263; eradication of, 123–24; food and, 124–26; I. G. Farben and, 232; as instigating the Dark Ages, 100; Liston and, 123–24; Martin and, 124–25; potato blight and, 25; rats and, 98, 100, 119–30; Simond and, 119–24, 127; typhus and, 89, 98

Flemming, Arthur, 263

flies: African sleeping sickness and, 68; Black Death and, 119; *Calliphora vomitoria* and, 192–93; chemicals of war and, 156, *157*, 192–93, 203, 205, 207, 211; DDT and, 185, 192–94, 200–208, 211, 251, 253–54, 276; ecological issues and, 251, 253–54, 256, 276; maggots and, 14; malaria and, 36, 42; sand, 256; trypanosomes and, 36; tsetse, 68; yellow fever and, 68, 74

Flitgun, 203, *210*

Florence, 102, 107

fluorine, x

fluoroethyl alcohol, 244–45

Fly Abatement Unit, 207

folidol, 246

food, 315; Black Death and, 98, 102–3, 106, 108, 124–26; concentration camps and, 222–23; DDT and, 194–95, 197, 206, 211, 214, 259; defoliants and, 288–91; ecological issues and, 249, 259, 263–65, 269–71, 274, 276, 280, 283, 286–92; fat content and, 231; fleas and, 124–26; I. G. Farben slaves and, 222; nutrition and, 103, 153, 274; organic, 274; pesticides and, 3, 26, 167–68, 194–95, 197, 206, 211, 214, 259, 263, 271, 288; potatoes and, 3, 5–6, 8, 26, 245; rats and,

98, 102–3; safety of, 249, 259, 263–65, 269–71, 274, 276, 280, 283, 286, 288–89, 291–92

Food Protection Committee, 274

Forbes, Stephen A., 156, 162

Ford, Gerald, 292–93

Ford's Theatre, 37

formaldehyde, 25, 59

formalin, 129

Fortune (magazine), 215–16

Fosse, Baron Chatry de la, 23

Franck, Dagmar, 300–309, 315

Franck, James: Berkeley lecture of, 300; Bohr and, 174, 176; death of, 316; Einstein and, 147, 301, 314–16; gas and, *148*, 149, 299; Haber and, 147–49, 172–74, 176, 178, 299, 304; Heidelberg doctorate and, 316; humanitarian relief and, 314–15; I. G. Farben and, 304; Max Planck Medal and, 316; Nazis and, 172, 174–76, 178, 299, 301, 304; Nobel Prize of, 74–75, 147, 149, 172, 174–75, 299, 303, 307–8; resignation of, 172, 299, 302–5; Teller and, 308; World War I and, 302, 309; World War II and, *309*

"Franck Report," 312–14

Franco-Prussian War, 17, 22, 35

Frank, Anne, 97

Frank, Margot, 97

Franklin, Benjamin, 5, 61

Free African Society, 54–55, 62

French, John, 147

French Academy of Sciences, 16, 36, 94

Freon, x, 191, 201–2, 297

Freud, Sigmund, 317

Fritsch, Karl, 179

Fritz Haber Institute, 183

frogs, 14, 105, 108

fuel, ix–xi, 207, 222, 289

German Corporation for Pest Control (Degesch) and, 167–69, 180, 194; increased use of, 295; Laveran and, 36; malaria and, 36, 49, 51; modern tragedies from, 287–90, 295–97; mortality of, 282; mosquitoes and, 49, 51, 81, 87, 198, 251, 253, 271, 288; persistent toxicity of, 251–53; potato blight and, 3, 23, 25–26; public opinion on, 263; resistance to, 212, 253–54, 256, 281, 288, 296–97; Schrader and, 243–46; specificity and, 252; systemic, 244–45; Technical Committee for Pest Control and, 167; trypanosomes and, 36; typhus and, 92; warnings over, 254–56; World War I and, 130, 160, 200, 271; World War II and, 246, 251, 282; yellow fever and, 81–82, 87. *See also specific chemicals*

Petrarch, Francesco, 100–104, 107, 126

pharmaceutical industry, 35, 264

phenol, 128–29, 129

Philadelphia, 7, 9, 53–62, 65, 76

Philippines, 184–85

phocomelia, 265

photocells, 310

photosynthesis, 308, 311

Phylloxera, 21–22, 245

physicians: Blackburn, 59–60; Black Death and, 107, 109–10, 113, 115, 117, 120, 127, 199–204; Brandt, 178; chemicals of war and, 178, 202, 222, 237; ecological issues and, 265; Finlay, 65–68, 76–77, 80, 87; Hippocrates, 31; Ho Kai, 109; King, 38; Lind, 91–92; Lowson, 109–10, 113, 115, 117; malaria and, 31–32, 37, 44, 51; Manson, 37; Millardet, 22; Nazis

and, 178, 222, 237; Nicolle, 92–97, 166; potato blight and, 11; Redi, 14; Reed, 67 (*see also* Reed, Walter); Ross, 51 (*see also* Ross, Ronald); Rush, 53, 55–57, 61–62, 65; Sammonicus, 32; Simond, 127, 199–204; spontaneous generation and, 14; typhus and, 91–97; yellow fever and, 53–87; Yersin, 112–19, 121–22, 126

Phytophthora infestans, 18, 20–22, 24–26

pigeons, 44, 113, 149

pitch, 91, 133

Planck, Erwin, 312

Planck, Max, 178, 301, 304, 309, 312, 316

plankton, 255

plastics, 218, 297, 310

plutonium, 314

pneumonia, 35

Poland, 178, 182, 219, 229, 230

Polányi, John, 303

Polányi, Michael, 303

polio, x, 86, 206

Pontine Marshes, 190

Pope, William J., 150, 177

Porter, William N., 225, 237

potato blight: *American Agriculturist* and, 272; Belgium and, 7; Berkeley and, 12–15, 18–19; Bourdeaux Mixture and, 24–26; De Bary and, 18, 20–22, 26; doctors and, 9–10; dry rot and, 7, 25; experiments and, 4, 16; famine and, 3–27, 88, 270–72; fleas and, 25; fungi and, 5, 7, 12–14, 18–20, 23, 25–26; globalization and, 3, 5, 9; guano and, 7, 12; immunity and, 10; inoculations and, 14, 18; insects and, 3, 9, 11, 21, 25; Irish and, xi,

Shaw, George Bernard, 249
Shawn, William, 268
sheep, 68, 110–11, 270, 293
Shell, 244, 257, 272
Sherwin-Williams, 205
ship fever. *See* typhus
Shute, Nevil, 278
Sibert, William, 160
Sicily, 101–2
Sierra Leone, 127
Silent Spring (Carson): as bestseller, 279; Book-of-the-Month Club and, 260, 269; chemical hazards after publication of, 287–98; chemical industry's reaction to, 271–76; as classic, 284; commercial success of, 269, 279; controversy over, 268–69, 318; dedication of, 283; ecological issues and, xi, 262, 264, 267–84, 287–90, 296–97, 311, 318; "A Fable for Tomorrow" and, 270–71; magazine reviews of, 275–78; potential titles for, 267–68; prepublication copies of, 268; reaction to by gender, 277–78; vetted facts of, 268
Silk Road, 101
Simmons, James Stevens, 184, 195–96, 198, 200–201, 212, 250–51
Simon & Schuster, 258
Simond, Paul-Louis, 119–24, 127
Singapore, 185
Skoda-Wetzler Works, 218
slavery, xi, 8; African, 3, 9, 32–34, 53–54, 90; Free African Society and, 54–55, 62; Haiti and, 58; I. G. Farben and, 220–25, 229, 231; immunity and, 33, 53; malaria and, 32–34, 53; typhus and, 90; yellow fever and, 33, 53–55, 58, 85
slugs, 14

smallpox, 68, 91, 112
Smith, Theobald, 67–68
smoke screens, 135–36
Smokey the Bear, 288–89
snakes, 14, 38, 185
Société Philomathique, 18, 20–22, 24–26
sodium aluminofluoride, 211
sodium fluoroacetate ("1080"), 213–14
soman, 238–39
Soviet Union, 220, *229*, 233, 237–40, 262, 315
Spain, 4, 33, 73–74, 90, 159
Spanish-American War, 67, 82, 184
spark plugs, ix
Sparta, 133–34
Speer, Albert, 233
spontaneous generation, 9, 12, 14–20, 26
sporangia, 19
Sports Illustrated (magazine), 278
squirrels, 160, 284
Sri Lanka, 214
SS (Schutzstaffel), 179–81, 221–23, *229*
SS John Harvey (ship), 235
stagnant water, 41, 49, 58, 66, 198, 205
Stalin, Joseph, 315
Stalingrad, 233
Standard Oil, ix
Stark, Johannes, 171–72
starvation, 7–10, 53, 98, 142, 158, 222, 271–72, 274
Stauffer, 244
steel, 150, 202, 230, 242, 293
Stegomyia fasciata mosquito, 65
stem rot, 25
stench, 11, 20, 72, 106
Stenhouse, 149
sterilization, 16, 72, 91, 284

von Laue, Max, 174–75, 177–78, 299
von Prowazek, Stanislaus Joseph Matthias, 95
von Wallenstein, Albrecht, 90
V-weapons, 232
VX nerve gas, 243–44, 293

Wake Island, 184
Wallace, Alfred Russel, 23–24
Wallace, Henry, 312
Walter Reed Hospital, 75
War Committee on Dielectrics, 311
war crimes, 209; doctors and, 190–91; Haber and, 163; I. G. Farben and, 223–25, 228, 233, 239, 242; Nernst and, 163; Nuremberg trials and, 182, 196, 215, 223–24, 228, 233, 242; Zyklon and, 181–82
War Crimes Investigation Unit, 181
War Production Board, 196, 201, 203, 311
Wasdin, Eugene, 79
Washington, George, 56
Washington Post, 71–72, 76, 269
water molds, 18–22, 26
weapons of mass destruction, 144, 293
weedkillers, 289–92
weevils, 25, 153
Weinbacher, Karl, 181
Weizmann, Chaim, 176
Westchester County, 257
Westing, Arthur, 292
Westinghouse, 201, 208
Whigs, 8
White, E. B., 261, 284
White Army, 97
White-Stevens, Robert, 275–76, 280, 286
Wilder, Russell Morse, 94
wildlife, 251, 254–55, 263, 275, 278, 282, 296

Willstätter, Richard, 141, 143, 149, 166, 177–78, 221, 238, 308
Wilson, Dr., 149
Wilson, Woodrow, 160
wine blight, 21–27
Wöhler, Friedrich, 137–38, 299
women, 257; birth defects and, 265, 290–92; Black Death and, 105–7; breast milk and, 272, 287; chemicals of war and, 157, 179, 222, 234; DDT and, 272, 287; ecological issues and, 269, 277–78, 283, 287; gunning down, 179; I. G. Farben and, 222; Nazis and, 179; stillbirths and, 287; thalidomide and, 264–65, 291
Women's National Book Association, 283
Women's Strike for Peace, 262
Woodham-Smith, Cecil, 5
Woods Hole, 258
World Health Organization, 288
World War I: chemicals of war and, 130, 133, 138, 142, 147–49, 156, 159–60, 166–70, 177–84, 188, 197, 199–200, 215–16, 235; ecological issues and, 249, 271; Franck and, 299, 302, 309; gas masks and, 299; malaria and, 33, 118; millionaires created by, 216; mortality of, 156–57; mustard gas and, 149–50, 157, 159, 169, 177; pesticides and, 130, 160, 200, 271; quinine and, 118, 188; trench fever and, 96, 127; typhus and, 96; von Hippel and, 301; yellow fever and, 83
World War II: chemicals of war and, 181, 188, 216, 219, 235, 246; Churchill and, 4–5, 136, 159, 199, 258, 279, 284, 305; ecological